PHYSICS
English–Chinese
Encyclopedic Dictionary
of general definitions, concepts and laws

物理学
英汉双解宝典
一般定义、概念及规律

[英]A. T. 奥古斯汀 A. T. Augousti	杨福玲 Yang Fuling	柳天宇 Liu Tianyu	主编 Edited
陈聪聪 Chen Congcong	谭 露 Tan Lu	张青滦 Zhang Qingluan	译 Translated

Physics: English-Ukrainian Encyclopedic Dictionary of general definitions, concepts and laws by Iryna Moroz, Copyright©2020 by Lviv Polytechnic Publishing House. All rights reserved.

This translation published under license. Simplified Chinese translation edition published by Tianjin University Press, Copyright©2023. All rights reserved.

版权合同：天津市版权局著作权合同登记图字02-2022-227号
本书简体中文版由利沃夫国立理工大学出版社授权天津大学出版社独家出版。

图书在版编目(CIP)数据

物理学英汉双解宝典：一般定义、概念及规律 / (英) A.T.奥古斯汀, 杨福玲, 柳天宇主编；陈聪聪, 谭露, 张青溓译. -- 天津：天津大学出版社, 2023.9
ISBN 978-7-5618-7408-0

Ⅰ.①物… Ⅱ.①A… ②杨… ③柳… ④陈… ⑤谭… ⑥张… Ⅲ.①物理学－普及读物－汉、英 Ⅳ.①O4-49

中国国家版本馆CIP数据核字(2023)第018762号

WULIXUE YINGHAN SHUANGJIE BAODIAN
YIBIAN DINGYI, GAINIAN JI GUILÜ

出版发行	天津大学出版社
地　　址	天津市卫津路92号天津大学内（邮编：300072）
电　　话	发行部：022-27403647
网　　址	www.tjupress.com.cn
印　　刷	北京盛通印刷股份有限公司
经　　销	全国各地新华书店
开　　本	710mm×1010mm　1/16
印　　张	22.25
字　　数	462千
版　　次	2023年9月第1版
印　　次	2023年9月第1次
定　　价	180.00元

凡购本书，如有缺页、倒页、脱页等质量问题，烦请与我社发行部门联系调换
版权所有　　侵权必究

Preface 前言

We are delighted to acknowledge the publication of *PHYSICS English-Chinese Encyclopedic Dictionary* which has been a multinational scholarly collaboration. It is based on an original *PHYSICS English-Ukrainian Encyclopedic Dictionary* that we developed and which was published in 2020, and we would like to thank the Chinese translation and editing team for their detailed and thorough work on this joint activity. We hope that it will find widespread usage and uptake among Chinese scholars who work in a wide variety of fields that use physics terminologies.

欣闻《物理学英汉双解宝典》即将出版，这是多国学者共同努力的结果！本书源自 2020 年我们共同出版的《物理学英语-乌克兰语百科图解词典》。我们谨对中国翻译和审校团队认真细致的工作表示由衷的感谢。希望本宝典能够在中国广为应用，成为各领域中国学者在运用物理学词条时的好助手。

Iryna Moroz
伊莉娜·莫拉兹
A. T. Augousti
A. T. 奥古斯汀

译者的话

《物理学英汉双解宝典》译自乌克兰利沃夫国立理工大学（Lviv Polytechnic National University）物理系副教授伊莉娜·莫拉兹（Iryna Moroz）编撰并与英国金斯顿大学（Kingston University）工程、计算机与环境学院教授 A. T. 奥古斯汀（A. T. Augousti）共同出版的《物理学英语-乌克兰语百科图解词典》。翻译团队由天津大学外国语学院和理学院的教师和学生组成，外国语学院杨福玲教授牵头翻译和审校工作，理学院柳天宇研究员负责物理学内容审核，外国语学院 2020 级研究生陈聪聪、谭露、张青滦参与翻译。在此谨对翻译团队全体人员在本书的翻译和审阅、修订过程中所做的工作表示衷心的感谢！

译者
2023 年 1 月

PHYSICS QUOTES	12
PHYSICS	14

MECHANICS   16

STANDARD of LENGTH, MASS, TIME	18
COORDINATE SYSTEM	20
SCALAR and VECTOR	22
SOME PROPERTIES of VECTOR	24
SCALAR PRODUCT	26
VECTOR PRODUCT	28
MOTION	30
POSITION, DISPLACEMENT, DISTANCE	32
VELOCITY and SPEED	34
ACCELERATION	36
FREELY FALLING OBJECT	38
PROJECTILE MOTION	40
UNIFORM CIRCULAR MOTION	42
FORCE	44
NEWTON'S FIRST LAW	46
NEWTON'S SECOND LAW	48
NEWTON'S THIRD LAW	50
WORK	52
ENERGY	54
CONSERVATION of ENERGY	56
LINEAR MOMENTUM	58
COLLISION	60
CENTER of MASS	62
STATIC EQUILIBRIUM	64
ROTATION of a RIGID OBJECT	66
ANGULAR VELOCITY and ACCELERATION	68
MOMENT of INERTIA	70
TORQUE	72
ROLLING MOTION	74
ANGULAR MOMENTUM and its CONSERVATION	76
NEWTON'S LAW of GRAVITY	78
KEPLER'S LAWS	80
ELASTIC PROPERTIES of SOLID	82
ELASTIC MODULUS	84
PRESSURE	86
FLUID STATICS	88
FLUID DYNAMICS	90

物理学名人名言	12
物理学	14

力学 16

长度、质量、时间的度量标准	18
坐标系	20
标量和矢量	22
矢量的性质	24
标量积	26
矢量积	28
运动	30
位置、位移、距离	32
速度和速率	34
加速度	36
自由落体	38
抛物运动	40
匀速圆周运动	42
力	44
牛顿第一定律	46
牛顿第二定律	48
牛顿第三定律	50
功	52
能	54
能量守恒	56
线性动量	58
碰撞	60
质心	62
静平衡	64
刚体的转动	66
角速度和角加速度	68
转动惯量	70
力矩	72
滚动	74
角动量和角动量守恒	76
牛顿万有引力定律	78
开普勒定律	80
固体的弹性	82
弹性模量	84
压强	86
流体静力学	88
流体动力学	90

VISCOSITY	92
SIMPLE HARMONIC MOTION	94
PENDULUM	96
DAMPED OSCILLATION	98
FORCED OSCILLATION	100
WAVE MOTION	102
INTERFERENCE of WAVE	104
SOUND WAVE	106
SHOCK WAVE	108
STATIONARY WAVE	110

THERMODYNAMICS 112

TEMPERATURE	114
THERMAL EXPANSION of SOLID	116
LAWS of an IDEAL GAS	118
ISOPROCESSES	120
HEAT	122
FIRST LAW of THERMODYNAMICS	124
HEAT TRANSFER	126
LAWS of RADIATION	128
MOLECULAR MODEL of an IDEAL GAS	130
EQUIPARTITION of ENERGY	132
BOLTZMANN DISTRIBUTION LAW	134
DISTRIBUTION of MOLECULAR SPEED	136
MEAN FREE PATH	138
VAN DER WAALS' EQUATION	140
SECOND LAW of THERMODYNAMICS	142
HEAT ENGINE and REFRIGERATOR	144
CARNOT ENGINE	146
GASOLINE ENGINE	148
ENTROPY	150

ELECTROMAGNETISM 152

ELECTRIC CHARGE	154
COULOMB'S LAW	156
ELECTRIC FIELD	158
ELECTRIC FLUX	160
GAUSS' LAW	162
ELECTROSTATIC EQUILIBRIUM	164
ELECTRIC POTENTIAL	166

黏性	92
简谐运动	94
摆	96
阻尼振动	98
受迫振动	100
波动	102
波的干涉	104
声波	106
冲击波	108
驻波	110

热力学   112

温度	114
固体热膨胀	116
理想气体定律	118
等温、等压和等容过程	120
热	122
热力学第一定律	124
热传递	126
辐射定律	128
理想气体分子模型	130
能量均分	132
玻尔兹曼分布律	134
分子速率分布	136
平均自由程	138
范德华方程	140
热力学第二定律	142
热机和制冷机	144
卡诺热机	146
汽油发动机	148
熵	150

电磁学   152

电荷	154
库仑定律	156
电场	158
电通量	160
高斯定律	162
静电平衡	164
电势	166

CAPACITANCE	168
CAPACITOR	170
COMBINATION of CAPACITORS	172
ELECTRIC DIPOLE and POLARIZATION	174
ELECTRIC CURRENT	176
RESISTANCE and OHM'S LAW	178
RESISTANCE and TEMPERATURE	180
ELECTROMOTIVE FORCE	182
COMBINATION of RESISTORS	184
KIRCHHOFF'S RULES	186
ELECTRICAL INSTRUMENT	188
MAGNETIC FIELD	190
MOTION of a CHARGED PARTICLE	192
CURRENT-CARRYING CONDUCTOR	194
HALL EFFECT	196
BIOT-SAVART LAW	198
AMPÈRE'S LAW	200
MAGNETIC FIELD of a SOLENOID	202
MAGNETIC FLUX	204
MAGNETISM in MATTER	206
FERROMAGNETISM	208
DIAMAGNETISM and PARAMAGNETISM	210
MAGNETIC FIELD of the EARTH	212
FARADAY'S LAW of ELECTROMAGNETIC INDUCTION	214
LENZ'S LAW	216
GENERATOR and MOTOR	218
EDDY CURRENT	220
MAXWELL'S EQUATIONS (I)	222
MAXWELL'S EQUATIONS (II)	224
INDUCTANCE	226
OSCILLATION in an LC CIRCUIT	228
RC, RL and RLC CIRCUITS	230
RESISTOR in an AC CIRCUIT	232
INDUCTOR in an AC CIRCUIT	234
CAPACITOR in an AC CIRCUIT	236
RLC SERIES CIRCUIT	238
TRANSFORMER	240
ELECTROMAGNETIC WAVE	242
ELECTROMAGNETIC SPECTRUM	244

电容	168
电容器	170
电容器的连接	172
电偶极子和电极化	174
电流	176
电阻和欧姆定律	178
电阻和温度	180
电动势	182
电阻器的连接	184
基尔霍夫定律	186
电气仪表	188
磁场	190
带电粒子的运动	192
通电导体	194
霍尔效应	196
毕奥-萨伐尔定律	198
安培定律	200
螺线管的磁场	202
磁通量	204
物质的磁性	206
铁磁性	208
抗磁性和顺磁性	210
地磁场	212
法拉第电磁感应定律	214
楞次定律	216
发电机和电动机	218
涡流	220
麦克斯韦方程组（Ⅰ）	222
麦克斯韦方程组（Ⅱ）	224
电感	226
LC电路中的振荡	228
RC电路、RL电路和RLC电路	230
交流电路中的电阻器	232
交流电路中的电感器	234
交流电路中的电容器	236
RLC串联电路	238
变压器	240
电磁波	242
电磁波谱	244

OPTICS 246

NATURE of LIGHT	248
REFLECTION of LIGHT	250
REFRACTION of LIGHT	252
DISPERSION	254
HUYGENS' PRINCIPLE	256
TOTAL INTERNAL REFLECTION	258
IMAGES FORMED by MIRRORS	260
THIN LENS	262
LENS ABERRATION	264
OPTICAL INSTRUMENT	266
INTERFERENCE of LIGHT WAVES	268
YOUNG'S DOUBLE-SLIT EXPERIMENT	270
INTERFERENCE in THIN FILM	272
NEWTON'S RINGS	274
MICHELSON INTERFEROMETER	276
DIFFRACTION	278
DIFFRACTION GRATING	280
DIFFRACTION of X-RAY by CRYSTAL	282
POLARIZATION of LIGHT WAVE	284
MALUS' LAW	286
BIREFRINGENCE	288
BREWSTER'S LAW	290

APPENDICES 292

GREEK ALPHABET and its COMMON USAGE	294
CHEMICAL SYMBOLS for the ELEMENTS	296
SELECTED FUNDAMENTAL CONSTANTS	300
PHYSICAL QUANTITIES	302
SI BASE UNITS and DERIVED UNITS	306
SELECTED PREFIXES for POWERS of TEN	307
MATHEMATICAL SYMBOLS and OPERATIONS	308
ALGEBRA	309
GEOMETRY	310
TRIGONOMETRY	311
ENGLISH INDEX	314
CHINESE INDEX	336

光学 246

- 光的本质 248
- 光的反射 250
- 光的折射 252
- 色散 254
- 惠更斯原理 256
- 全反射 258
- 镜面成像 260
- 薄透镜 262
- 透镜像差 264
- 光学仪器 266
- 光波干涉 268
- 杨氏双缝干涉实验 270
- 薄膜干涉 272
- 牛顿环 274
- 迈克尔逊干涉仪 276
- 衍射 278
- 衍射光栅 280
- 晶体的X射线衍射 282
- 光的偏振 284
- 马吕斯定律 286
- 双折射 288
- 布儒斯特定律 290

附录 292

- 希腊字母表及常见用法 294
- 化学元素符号 296
- 常用基本常数 300
- 物理量 302
- 国际单位制基本单位和导出单位 306
- 常用10的幂次前缀 307
- 数学符号和运算 308
- 代数 309
- 几何 310
- 三角函数 311
- 英文索引 314
- 中文索引 336

PHYSICS QUOTES
物理学名人名言

Nature has a great simplicity
and therefore
a great beauty.
Richard Feynman

大自然因至简而至美。
理查德·费曼

To explain all nature
is too difficult a task
for any one man or even
for any one age.
Isaac Newton

对于任何一个人抑或一个时代而言，要完全解释清楚大自然都是一项难以完成的艰巨任务。
艾萨克·牛顿

Physics
is the ultimate intellectual adventure,
the quest to understand
the deepest mysteries
of our Universe.
Physics doesn't take something
fascinating and make it boring.
Rather, it helps us see more clearly,
adding to the beauty
and wonder of the world around us.
Max Tegmark

物理学是终极的智力冒险，是对理解宇宙最深奥秘的探索。物理学不会把有趣的东西变得无聊，相反，物理学通过给我们身边的世界增添魅力和奇观让我们把世界看得更加清晰。
马克斯·泰格马克

Mathematics began to seem too
much like puzzle solving.
Physics is puzzle solving, too,
but of puzzles created by nature,
not by the mind of man.
Maria Goeppert-Mayer

数学越来越像解谜游戏。物理也在解谜，只不过谜题出自大自然而非人为。
玛丽亚·格佩特-梅耶

One must divide one's time
between politics and equations,
because politics is for the present,
while our equations
are for eternity.
Albert Einstein

一个人应该把时间花在政治和解方程上，因为政治代表当下，方程代表永恒。
阿尔伯特·爱因斯坦

The laws of physics is the canvas
God laid down
on which to paint his masterpiece.
Dan Brown

如果世界是上帝创作的不朽画作，那么物理学定律就是这幅画作的画布。
丹·布朗

A mathematician
may say anything he pleases,
but a physicist
must be at least partially sane.
Josiah Williard Gibbs

数学家想说什么就说什么，但物理学家至少要保持部分理智。
约西亚·威利亚德·吉布斯

I think nature's imagination
is so much greater than man's,
she's never going to let us
relax.
Richard Feynman

我认为大自然的想象力远比人类丰富得多，她永远都不会让我们闲着。
理查德·费曼

We have sought for firm ground
and found none.
The deeper we penetrate, the more
restless becomes the universe;
all is rushing about and vibrating
in a wild dance.
Max Born

我们一直在探求坚实的理论基础，但却一无所获。我们研究得越深，宇宙就看似越无章法，一切都好像在狂舞中奔忙着、振动着。
马克斯·玻恩

There is no science in this world
like physics.
Nothing comes close to the precision
with which physics enables you
to understand the world around you.
Neil deGrasse Tyson

在这个世界上，物理学独一无二，因为没有任何其他科学能像物理学那样让你能精准地理解周围的世界。
尼尔·德格拉斯·泰森

Physics
is mathematical not because
we know so much
about the physical world,
but because we know so little;
it is only its mathematical properties
that we can discover.
Bertrand Russell

物理学具有数学特性不是因为我们对物理世界知道得太多，而是因为我们知道得太少；我们能发现的只是它的数学特性。
伯特兰·罗素

PHYSICS
物理学

TERMS	术语
physics	物理学
classical physics	经典物理学
modern physics	现代物理学
mechanics	力学
thermodynamics	热力学
electromagnetism	电磁学
optics	光学
atomic physics	原子物理学
nuclear physics	核物理学
condensed matter physics	凝聚态物理学
physics of elementary particles	基本粒子物理学
high energy physics	高能物理学
theory of relativity	相对论
quantum mechanics	量子力学
theory	理论
concept	概念
law	定律
hypothesis	假设
experiment	实验
physical phenomenon	物理现象
physical quantity	物理量
experimental observation	实验观测
quantitative measurement	定量测量

LANGUAGE OF PHYSICS / 物理学用语

Physics — science that deals with the structure of matter and the interactions between the fundamental constituents of the observable universe. Physics is concerned with all aspects of nature on both the macroscopic and sub-microscopic levels. Its scope of study encompasses not only the behaviour of objects under the action of given forces but also the nature and origin of gravitational, electromagnetic, and nuclear force fields. Its ultimate objective is the formulation of a few comprehensive principles that bring together and explain all such disparate phenomena.

(*Encyclopedia Britannica*)

物理学：研究物质结构和可观测的宇宙基本成分之间相互作用的科学。物理学从宏观和亚微观的角度研究自然界的方方面面。其研究范围不仅包括物体受力后的行为，而且还包括引力场、电磁场和核力场的本质和起源。物理学的终极目标是创建几个能够整合和解释各种现象的综合原理。

《大英百科全书》

MECHANICS

On February 6, 2018, the Falcon Heavy was launched, carrying Elon Musk's Tesla Roadster into space. The launch and operation of a space rocket involves many fundamental principles of classical mechanics, thermodynamics, and electromagnetism.

2018年2月6日,"重型猎鹰"火箭发射升空,搭载着埃隆·马斯克的特斯拉跑车进入太空。火箭的发射和运行涉及经典力学、热力学和电磁学的诸多基本原理。

力学

MECHANICS

Mechanics is a branch of physics concerned with the behaviour of bodies under the action of forces. This includes both the movement and the deformation of bodies that may be at rest or in motion.

力学

力学：物理学的一个分支，研究在力的作用下物体的行为，包括静止或运动物体的运动和形变。

BRANCHES OF MECHANICS

Kinematics is a branch of mechanics that describes the motion of a body without regard to the cause of that motion.

Dynamics is a branch of mechanics concerned with the relationship between forces and the motion of bodies arising from these forces.

Statics is a branch of mechanics that deals with bodies in equilibrium.

力学分支

运动学：力学的一个分支，研究一个物体的运动而不涉及运动的起因。

动力学：力学的一个分支，研究力和在力的作用下物体所发生的运动之间的关系。

静力学：力学的一个分支，研究平衡状态下的物体。

STANDARD of LENGTH, MASS, TIME
长度、质量、时间的度量标准

TERMS 术语

standard of length	长度标准
standard of mass	质量标准
standard of time	时间标准
International System of Units	国际单位制
(SI — Système Internationale)	（SI——国际单位制）
CGS (centimeter-gram-second) system	厘米-克-秒制
foot-pound-second system	英尺-磅-秒制
SI unit	国际单位制单位
meter (m)	米
kilogram (kg)	千克
second (s)	秒
conversion factor	换算因子

LANGUAGE OF PHYSICS 物理学用语

Length, **mass** and **time** are three basic quantities in mechanics. All other physical quantities in mechanics can be expressed in terms of these three.

长度、质量和时间：力学的三个基本量。力学中的其他所有物理量都可以用这三个量来表示。

The **meter** is defined as the distance traveled by light in a vacuum in a time of 1/299 792 458 s.

米：光在真空中经过 1/299 792 458 秒传播的距离。

The **kilogram** is defined as the mass of a specific platinum-iridium alloy cylinder.

千克：一个特制的铂铱合金圆柱体的质量。

The **second** is 9 192 631 770 cycles of radiation emitted by an excited atom of cesium-133.

秒：铯-133 原子受激辐射出的电磁波振荡 9 192 631 770 个周期所需要的时间。

The **International System of Units (SI)** is the internationally adopted system of units used by all the scientists and almost all countries of the world.

国际单位制：几乎所有国家和所有科学家采用的国际通用的单位系统。

A **conversion factor** is a factor by which a quantity expressed in one set of units must be multiplied in order to express that quantity in different units.

换算因子：同一物理量在不同单位下转换所需要乘的因子。

STANDARD OF MASS

The International Standard Kilogram was established in 1887 and kept at the International Bureau of Weights and Measures (Bureau International des Poids et Mesures) at Sevres, France. In 2018 the kilogram was defined by taking the fixed numerical value of the Planck constant h to be $6.626\ 070\ 15 \times 10^{-34}$ when expressed in the unit J·s, which is equal to kg·m²/s.

质量标准

国际标准千克于1887年被确立，并由法国塞夫勒的国际度量衡局保管。2018年，1千克被定义为对应普朗克常数h为$6.626\ 070\ 15 \times 10^{-34}$焦耳·秒（或者千克·米²/秒）时的质量单位。

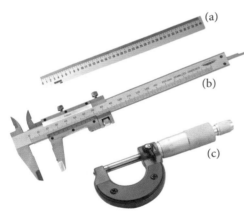

STANDARD OF LENGTH

Tools such as (a) a ruler, (b) a caliper and (c) a micrometer can be used for measuring length with different precision.

长度标准

直尺（a）、卡尺（b）、千分尺（c）等工具可以以不同精度来测量长度。

STANDARD OF TIME

An atomic clock. This device keeps time with an accuracy of about 3 millionths of a second per year.

时间标准

原子钟，该装置的计时误差约为每年百万分之三秒。

COORDINATE SYSTEM
坐标系

TERMS 术语

location in space	空间位置
position	位置
point on a line	直线上的点
point in a plane	平面内的点
point in space	空间中的点
coordinate	坐标
set of coordinates	坐标系
reference frame	参考系
reference point	参考点
reference axis	参考轴
origin	原点
axis	坐标轴
set of axes	坐标轴集
scale	刻度
label on the axis	坐标轴的名称
positive direction	正方向
negative direction	负方向
clockwise	顺时针
counterclockwise	逆时针
X-axis	X轴
Y-axis	Y轴
Z-axis	Z轴
polar axis	极轴
abscissa	横坐标
ordinate	纵坐标
unit vector i, j, k	单位矢量i、j、k
Cartesian coordinate system	笛卡儿坐标系
rectangular coordinate system	直角坐标系
polar coordinate system	极坐标系
cylindrical coordinate system	柱坐标系
spherical coordinate system	球坐标系

LANGUAGE OF PHYSICS   物理学用语

A **coordinate system** is a system for specifying the location of a point in space using coordinates measured in a specified way.

坐标系：按规定方法确定空间内一点位置所采用的坐标系统。

A coordinate system consists of an origin (a fixed reference point) and a set of specified axes with appropriate scales and labels on the axes.

坐标系包括一个原点（固定的参考点）和一组有适当刻度和名称的特定坐标轴。

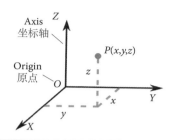
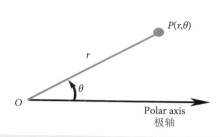

CARTESIAN COORDINATES

Designation of point in a Cartesian coordinate system. A point on a line can be described with one coordinate. A point in a plane is located with two coordinates, and a point in space is labelled with three coordinates (x, y, z).

笛卡儿坐标

笛卡儿坐标系中点的标记。直线上的一点可以用一个坐标来表示；平面内的点用两个坐标来表示；空间中的点用三个坐标（x，y，z）来表示。

POLAR COORDINATES

The plane polar coordinates of a point are represented by the distance r and the angle θ which is measured counterclockwise from the positive X-axis.

极坐标

平面内一点的极坐标由距离r和角度θ来表示，并且规定从X轴的正半轴起逆时针旋转的角度为正。

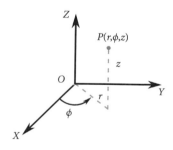
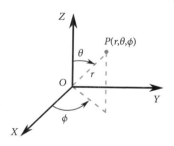

CYLINDRICAL COORDINATES

The cylindrical coordinates of a point are represented by the two distances r, z and the angle ϕ.

柱坐标

一个点的柱坐标由两个距离r、z和角度ϕ表示。

SPHERICAL COORDINATES

The spherical coordinates of a point are represented by the distance r and two angles ϕ, θ.

球坐标

一个点的球坐标由距离r和两个角ϕ、θ表示。

SCALAR and VECTOR
标量和矢量

TERMS 术语

scalar quantity	标量
vector quantity	矢量
magnitude	大小
direction	方向
tail	矢量起点
tip	矢量终点
number	数值
unit	单位
change in quantity	量的变化
boldface letter	粗体字母
arrow over the letter	字母上方的箭头
projection	投影
component	分量
unit vector	单位矢量
scalar product	标量积
dot product	点乘
vector product	矢量积
cross product	叉乘

LANGUAGE OF PHYSICS 物理学用语

A **scalar quantity** is a quantity that can be completely described by a magnitude, that is, by a number and, usually, a unit.

标量：一种可以完全用大小来描述的量，即用一个数值（通常有单位）来描述。

A **scalar quantity** has only magnitude and no direction.

标量只有大小，没有方向。

Examples of scalar quantities are mass, volume, time interval, speed, temperature, work, energy and distance.

标量的例子有质量、体积、时间间隔、速度、温度、功、能量和距离。

A **vector quantity** is a quantity that needs both a magnitude and a direction to completely describe it.

矢量：一个既需要大小又需要方向才能完全描述的量。

A **vector quantity** has both magnitude and direction.

矢量既有大小，又有方向。

Examples of vector quantities are force, displacement, velocity, acceleration, momentum and torque.

矢量的例子有力、位移、速度、加速度、动量和力矩。

VECTOR

To describe the vector completely, we must specify both the direction and a number to indicate the magnitude of the vector.

矢量

要完整地描述矢量，我们必须同时指定矢量的方向和表示矢量大小的数值。

VECTOR QUANTITY

A vector quantity is a physical quantity that is completely specified by a number with appropriate units plus a direction.

矢量

矢量是一个物理量，它由一个带有恰当单位的数值加上一个方向来完整表述。

SCALAR QUANTITY

One example of a scalar quantity is the number of apples.

标量

标量的一个例子是苹果的数量。

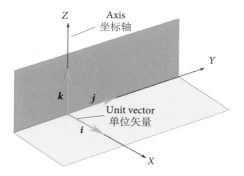

UNIT VECTORS

The unit vectors i, j, k are directed along the X, Y, Z axes respectively.

单位矢量

单位矢量i、j、k分别沿X、Y、Z轴方向。

SOME PROPERTIES of VECTOR
矢量的性质

TERMS	术语
equality of vectors	矢量相等
addition of vectors	矢量加法
resultant vector	合矢量
triangle method of addition	三角形加法法则
polygon method of addition	多边形加法法则
parallelogram method of addition	平行四边形加法法则
commutative law of addition	加法交换律
associative law of addition	加法结合律
negative of a vector	负矢量
subtraction of vectors	矢量减法
multiplication of a vector by a scalar	矢量和标量相乘
Pythagoras theorem	毕达哥拉斯定理
sine function	正弦函数
cosine function	余弦函数
tangent function	正切函数
diagonal of parallelogram	平行四边形的对角线
side of triangle	三角形的边

LANGUAGE OF PHYSICS
物理学用语

The **resultant vector** is the vector sum of any number of vectors.

合矢量：任意多个矢量的矢量和。

In the **polygon method** we add each vector to the preceding vector by placing the tail of one vector next to the head of the previous vector. The resultant vector is the sum of all these vectors.

在**多边形法则**中，我们通过将一个矢量的末端和前一个矢量的始端相连来实现矢量求和。合矢量就是所有这些矢量的和。

The **parallelogram method of vector addition** says that the main diagonal of a parallelogram is equal to the magnitude of the sum of the vectors that make up the sides of the parallelogram.

矢量加法的平行四边形法则表明平行四边形的主对角线的长度等于平行四边形两条邻边矢量之和的大小。

The **component of a vector** is the projection of a vector onto a specified axis.

矢量的分量：矢量在指定轴上的投影。

The **x, y or z-component of a vector** is the length of the projection of the vector onto the X, Y or Z-axis respectively.

矢量的x、y或z分量分别是该矢量在X、Y或Z轴上的投影长度。

TRIANGLE METHOD OF ADDITION

When vector **A** is added to vector **B**, the resultant **R** is the vector that runs from the tail of **A** to the tip of **B**. This method can be extended to the addition of many vectors, and is then known as the polygon method.

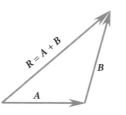
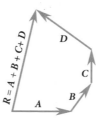

三角形加法法则

当矢量**A**与矢量**B**相加时，合矢量**R**就是连接**A**的起点和**B**的终点的矢量。这种方法可以扩展到多个矢量相加，也就是多边形法则。

PARALLELOGRAM METHOD OF ADDITION

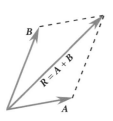

The resultant vector **R** is the diagonal of a parallelogram with sides **A** and **B**.

平行四边形加法法则

合矢量**R**就是以**A**和**B**为邻边的平行四边形的对角线。

COMMUTATIVE LAW OF ADDITION

Geometric construction for verifying the commutative law of addition.

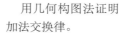
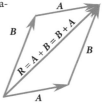

加法交换律

用几何构图法证明加法交换律。

ASSOCIATIVE LAW OF ADDITION

Geometric construction for verifying the associative law of addition.

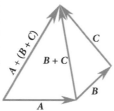
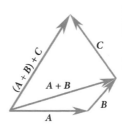

加法结合律

用几何构图法证明加法结合律。

SUBTRACTION OF VECTORS

The operation **A−B** is defined as vector **−B** added to vector **A**.

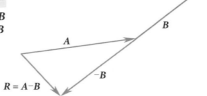

矢量减法

运算**A−B**即矢量**−B**加矢量**A**。

SCALAR PRODUCT
标量积

TERMS	术语
scalar product	标量积
dot symbol	点乘符号
dot product	点积
commutative law of multiplication	乘法交换律
distributive law of multiplication	乘法分配律
unit vector	单位矢量
magnitude	大小
scalar projection (scalar component)	标量投影（标量分量）
angle	角度
cosine	余弦

LANGUAGE OF PHYSICS / 物理学用语

The **scalar product** of any two vectors A and B is a scalar quantity equal to the product of the magnitudes of the two vectors and the cosine of the angle θ between them:

$$A \cdot B = AB\cos\theta$$

Shorthand notation for $AB\cos\theta$ is

$$A \cdot B$$

(read "A dot B")

The scalar product obeys the commutative and distributive laws of multiplication. The scalar product is a mathematical operation which combines two vectors and produces a scalar as a result. It is widely used in many areas of physics.

任意两个矢量A和B的**标量积**等于这两个矢量的大小乘以它们之间夹角θ的余弦值：

$$A \cdot B = AB\cos\theta$$

$AB\cos\theta$简写为

$$A \cdot B$$

（读作A点乘B）

标量积符合乘法交换律和分配律。标量积是一种数学运算，它将两个矢量组合而产生一个标量。标量积广泛应用于物理学的诸多领域。

SCALAR PRODUCT

The scalar product $A \cdot B$ equals the magnitude of A multiplied by the projection of B onto A. This expression is symmetrical: it is also equal to the magnitude of B multiplied by the projection of A onto B.

标量积

标量积 $A \cdot B$ 等于 A 的大小乘以 B 在 A 上的投影。这个表达式是对称的：它也等于 B 的大小乘以 A 在 B 上的投影。

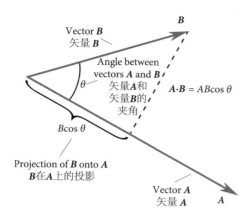

DOT PRODUCT OF UNIT VECTORS

These equations follow from the definition of the unit vectors, and are true for any orthogonal coordinate system.

单位矢量的点乘

这些等式源于单位矢量的定义，并且对任意正交的坐标系都成立。

$$i \cdot i = j \cdot j = k \cdot k = 1$$
$$i \cdot j = i \cdot k = j \cdot k = 0$$

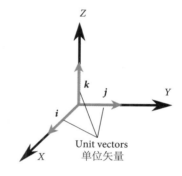

WORK AS A DOT PRODUCT

The work done by a constant force can be expressed in terms of the scalar product of the force vector and the displacement vector.

用点积表示功

恒定力所做的功可以用力矢量和位移矢量的标量积来表示。

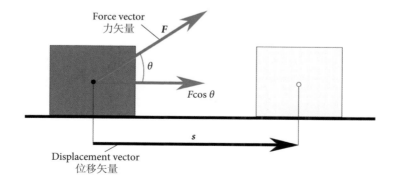

VECTOR PRODUCT
矢量积

TERMS 术语

vector product	矢量积
cross symbol	叉乘符号
cross product	叉积
parallelogram area	平行四边形面积
perpendicular	垂直的
right-hand rule	右手定则
direction	方向
plane formed by vectors	矢量构成的平面

LANGUAGE OF PHYSICS 物理学用语

The **vector** (or **cross**) **product** of any two vectors A and B is a third vector C, the magnitude of which is $AB\sin\theta$, where θ is the angle included between A and B and whose direction is at right angles to the plane formed by the vectors, in a direction defined by the right-hand rule:

$$C = A \times B$$

$$C = AB\sin\theta$$

Shorthand notation for $AB\sin\theta$ is

$$A \times B$$

(read "A cross B")

The vector (or cross) product is a mathematical operation which combines two vectors and produces a third vector as a result. It is widely used in many areas of physics.

The **vector product of unit vectors** is defined as

$$i \times j = -j \times i = k$$
$$i \times k = -k \times i = j$$
$$j \times k = -k \times j = i$$

任意两个矢量 A 和 B 的**矢量积**（或**叉积**）为一矢量 C。C 的大小为 $AB\sin\theta$，其中 θ 为 A 和 B 间的夹角。C 的方向垂直于 A 和 B 构成的平面，并且符合右手定则：

$$C = A \times B$$

$$C = AB\sin\theta$$

$AB\sin\theta$ 简写为

$$A \times B$$

（读作 A 叉乘 B）

矢量积（或叉积）是一种由两个矢量构成第三个矢量的数学运算，广泛应用于物理学的诸多领域。

单位矢量的矢量积定义为

$$i \times j = -j \times i = k$$
$$i \times k = -k \times i = j$$
$$j \times k = -k \times j = i$$

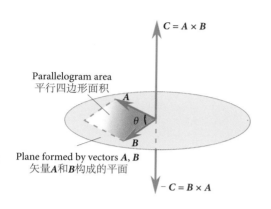

VECTOR PRODUCT

The vector product $A \times B$ is a third vector C having a magnitude $AB\sin\theta$ equal to the area of the parallelogram shown. The direction of C is perpendicular to the plane formed by A and B, and its sense is determined by the right-hand rule.

矢量积

矢量积 $A \times B$ 构成第三个矢量 C，其大小为 $AB\sin\theta$，即如图所示平行四边形面积。C 的方向垂直于 A 和 B 构成的平面，并且符合右手定则。

RIGHT-HAND RULE

The four fingers of the right hand are pointed along A and then "wrapped" into B through angle θ. The direction of the erect right thumb is the direction of $A \times B$.

右手定则

右手的四个手指指向 A 的方向，然后通过 θ 角卷向 B。右手大拇指竖直的方向就是 $A \times B$ 的方向。

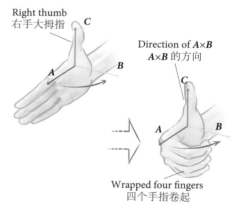

TORQUE AS A VECTOR PRODUCT

The torque vector τ lies in a direction perpendicular to the plane formed by the position vector r and the applied force F.

用矢量积表示力矩

力矩矢量 τ 的方向垂直于位置矢量 r 和作用力 F 所构成的平面。

MOTION
运动

TERMS	术语
kinematics	运动学
motion	运动
translational motion	平动
rotational motion	转动
vibrational (or oscillatory) motion	振动
one-dimensional motion	一维运动
two-dimensional motion	二维运动
motion in a plane	平面内的运动
motion along a straight line	直线运动
clockwise motion	顺时针运动
counterclockwise (or anticlockwise) motion	逆时针运动
moving object	运动物体
space and time	空间和时间
particle	质点
body	物体
object	物体
particle model	质点模型
linear motion	直线运动
curvilinear motion	曲线运动
random motion	随机运动
Brownian motion	布朗运动
circular motion	圆周运动
rolling motion	滚动
projectile motion	抛物运动
simple harmonic motion	简谐运动
oscillation	振荡
oscillatory motion	振动
wave motion	波动
relative motion	相对运动

LANGUAGE OF PHYSICS 物理学用语

Motion is a change in the position of an object over time.

运动：物体位置随时间的变化。

The **particle model** means treating a moving object as a point having no size.

质点模型把一个运动的物体看作一个没有大小的点。

Classical mechanics is used for describing the motion of macroscopic objects. Quantum mechanics describes physical reality at the atomic level of matter.

经典力学用于描述宏观物体的运动。量子力学则用于描述物质在原子尺度上的物理特性。

力学 31

TRANSLATIONAL MOTION
A car moving along a highway is undergoing translational motion.

平动
沿高速公路行驶的汽车正在进行平动。

MOTION
Motion represents the continuous change in the position of an object.

振动
单摆的往复运动就是振动的一个例子。

运动
运动表示物体位置的连续变化。

OSCILLATORY MOTION
The back-and-forth motion of a pendulum is an example of oscillatory motion.

ROTATIONAL MOTION
The Earth's daily spin on its axis is an example of rotational motion.

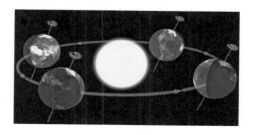

转动
地球绕地轴每日自转就是转动的一个例子。

POSITION, DISPLACEMENT, DISTANCE
位置、位移、距离

TERMS	术语
position	位置
initial position	初始位置
final position	最终位置
position vector	位置矢量
change in position	位置变化
Greek letter delta Δ	希腊字母德尔塔 Δ
time interval	时间间隔
function of time	时间函数
position-time graph	位置-时间图
slope	斜率
path	路径
length of a path	路径长度
displacement	位移
displacement vector	位移矢量
positive displacement	正位移
negative displacement	负位移
zero displacement	零位移
distance	距离

LANGUAGE OF PHYSICS　　　　　　　　物理学用语

A particle's **position** is the location of the particle with respect to a chosen reference point.

质点的**位置**：该点相对于所选参考点的位置。

The **position vector** of a particle is drawn from the origin of some reference frame to the particle located in an appropriate coordinate system.

一个质点的**位置矢量**是在一个适当的坐标系中由某个参考系的原点引向质点所在位置的矢量。

The **displacement** is the change in the position of the particle in some time interval.

位移：质点在某段时间间隔内位置的变化。

The **displacement vector** for a particle is the difference between its final position vector and its initial position vector.

一个质点的**位移矢量**就是它的最终位置矢量和它的初始位置矢量的差。

Distance is the length of a path followed by a particle.

距离：一个质点运动路径的长度。

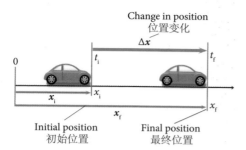

POSITION VECTOR

The point whose Cartesian coordinates are (x, y) can be represented by the position vector $r = x\mathbf{i} + y\mathbf{j}$.

位置矢量

笛卡儿坐标为 (x, y) 的点可以用位置矢量 $r = x\mathbf{i} + y\mathbf{j}$ 表示。

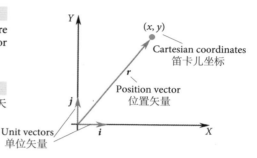

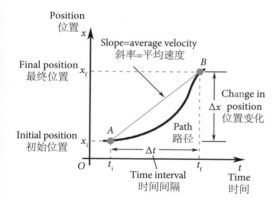

POSITION-TIME GRAPH

The slope of the line on a position-time graph reveals useful information about the velocity of the object.

位置-时间图

位置-时间图上曲线的斜率揭示了有关物体速度的有用信息。

Difference between displacement and distance.

位移和距离的区别。

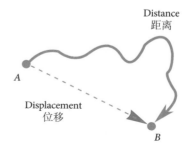

VELOCITY and SPEED
速度和速率

TERMS	术语
velocity	速度
constant velocity	匀速
average velocity	平均速度
instantaneous velocity	瞬时速度
relative velocity	相对速度
positive velocity	正速度
negative velocity	负速度
zero velocity	零速度
speed	速率
magnitude	大小
direction	方向
vector quantity	矢量
scalar quantity	标量
position-time graph	位置-时间图
slope of line	曲线斜率
tangent line	切线
limit of the ratio	比值的极限

LANGUAGE OF PHYSICS
物理学用语

The **speed** of a particle is equal to the magnitude of its velocity.

质点的**速率**等于其速度的大小。

The **average speed** is the distance that a body moves per unit time.

平均速率：物体在单位时间内移动的距离。

The **average velocity** is the average rate at which the displacement (i.e. position) vector changes with time.

平均速度：位移（即位置）矢量随时间变化的平均速率。

The **instantaneous velocity** of a particle is the limit of the ratio of its displacement and the time interval as the time interval approaches zero.

质点的**瞬时速度**是时间间隔趋于零时其位移与时间间隔比值的极限。

The **instantaneous velocity** equals the derivative of the position vector with respect to time.

瞬时速度等于位置矢量对时间的导数。

When an object travels the same distance every second, then the object is said to be moving with **constant speed**.

当一个物体每秒运动相同的距离时，称该物体在做**匀速**运动。

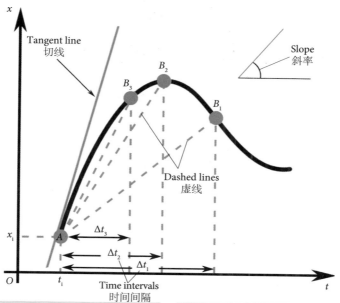

VELOCITY FROM A POSITION-TIME GRAPH

Position-time graph for the particle. As the time intervals get smaller and smaller, the average velocity for that interval, equals to the slope of the dashed line connecting A and the appropriate B, approaching the slope of the tangent line at B. The instantaneous velocity at A is the slope of the blue tangent line at time t_i.

根据位置-时间图求速度

质点的位置-时间图。随着时间间隔越来越小，这个时间间隔内的平均速度等于连接 A 点和相应 B 点的虚线的斜率，且约等于 B 点切线的斜率。A 点的瞬时速度等于 t_i 时刻蓝色切线的斜率。

SIGN OF VELOCITY

In this position-time graph, the velocity is positive at A, where the slope of the tangent line is positive, zero at B, where the slope of the tangent line is zero, and negative at C, where the slope of the tangent line is negative.

速度的正负

在该位置-时间图中，速度在 A 点是正的，因为该点处切线斜率为正；在 B 点为零，因为切线斜率为零；在 C 点是负的，因为切线斜率为负。

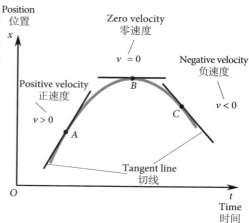

ACCELERATION
加速度

TERMS 术语

acceleration	加速度
average acceleration	平均加速度
instantaneous acceleration	瞬时加速度
constant acceleration	恒定（匀）加速度
relative acceleration	相对加速度
positive acceleration	正加速度
negative acceleration	负加速度
zero acceleration	零加速度
velocity-time graph	速度-时间图
slope of the graph	图像斜率
time interval	时间间隔
initial velocity	初速度
final velocity	末速度
limit	极限

LANGUAGE OF PHYSICS 物理学用语

Acceleration is the rate at which the velocity of a moving body changes with time.

加速度：运动物体的速度随时间的变化率。

When an accelerating object changes its velocity by the same amount each second, then the object is said to be moving with **constant acceleration**.

当一个加速运动的物体每秒速度的变化量相同时，称该物体在做**匀加速度**运动。

The **average acceleration** of a particle during some time interval is defined as the ratio of the change in its velocity and the time interval.

质点在某一时间间隔内的**平均加速度**是其速度变化量与该时间间隔之比。

The **instantaneous acceleration** is the limit of the ratio of the change in its velocity and the time interval as the time interval approaches zero.

瞬时加速度：当时间间隔趋于零时，其速度变化量与时间间隔比值的极限。

The **instantaneous acceleration** equals the derivative of the velocity vector with respect to time.

瞬时加速度等于速度矢量对时间的导数。

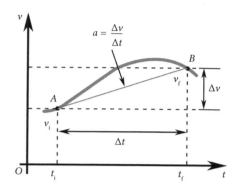

AVERAGE ACCELERATION

Velocity-time graph for a particle moving in a straight line. The slope of the blue straight line connecting A and B is the average acceleration in the time interval Δt.

平均加速度

质点沿直线运动的速度-时间图。连接A点和B点的蓝色直线的斜率是时间间隔Δt内的平均加速度。

Top fuel dragsters are the quickest accelerating racing cars in the world with the fastest competitors reaching speeds of 539 km/h and finishing the 402 m runs in 4.5-4.8 s. They accelerate from a standstill to 160 km/h in 0.8 s.

顶级燃料短程高速赛车是世界上加速最快的赛车。最快的参赛者的车速可达539千米/小时，可在4.5~4.8秒内跑完402米。这种车能在0.8秒内从静止加速到160千米/小时。

MOTION WITH CONSTANT ACCELERATION

A particle moving along the X-axis with constant acceleration $\boldsymbol{a}$: (a) the acceleration-time graph; (b) the velocity-time graph; (c) the position-time graph.

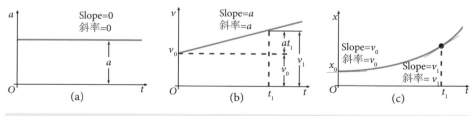

匀加速运动

一个质点以恒定加速度$\boldsymbol{a}$沿X轴运动：(a)加速度-时间图；(b)速度-时间图；(c)位置-时间图。

FREELY FALLING OBJECT
自由落体

TERMS 术语	
free-fall	自由落体
free-fall motion	自由落体运动
free-fall acceleration	自由落体加速度
acceleration due to gravity	重力加速度
gravity	重力
Earth's gravity	地球引力
gravitation	万有引力
Newton's theory of gravitation	牛顿的万有引力理论
gravitational field	重力（引力）场
uniform gravitational field without air resistance	忽略空气阻力的均匀重力场
uniform gravitational field with air resistance	考虑空气阻力的均匀重力场
gravitational force	引力
weight	重量
weightlessness	失重

LANGUAGE OF PHYSICS   物理学用语

A **freely falling body** is any body that is moving under the influence of gravity only or any body that is dropped or thrown near the surface of the Earth.

自由落体：仅在重力作用下运动的物体，或在地球表面附近掉落或抛掷的物体。

If air friction is neglected, all objects that are dropped near the surface of the Earth, are accelerated toward the center of the Earth with an acceleration of approximately $g = 9.8$ m/s².

如果忽略空气摩擦，所有在地球表面附近坠落的物体都加速向地球中心运动，加速度约为 $g = 9.8$ 米/秒²。

Even on the surface of the Earth, there are local **variations in the value of the acceleration due to gravity**. These variations are due to latitude, altitude and the local geological structure of the region.

即使在地球表面，**由于重力的作用，加速度的值也会有局部的变化**。其影响因素包括该地区的纬度、海拔和局部地质结构。

Weightlessness is an absence of stress and strain resulting from externally applied mechanical contact-forces, typically normal forces.

失重：由于外部施加的机械接触力（通常是法向力）而导致的应力和应变的缺失。

力学 39

FALLING FEATHER AND APPLE

The apple and feather fall almost together when the air resistance is negligible. In real conditions the apple will reach the ground first. The feather experiences greater air resistance resulting in a smaller net force acting upon it and it falls more slowly than the apple.

下落的羽毛和苹果

若忽略空气阻力，苹果和羽毛几乎同时落下。但在实际情况下，苹果会先着地。羽毛受到的空气阻力更大，作用在羽毛上的净力就更小，因此羽毛比苹果下落得更慢。

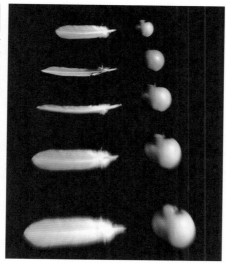

Leonid Kadenyuk is the first astronaut of independent Ukraine. He made his flight on NASA's Columbia Space Shuttle in 1997 as part of the international mission STS-87. He is shown here with the symbols of Ukraine during weightlessness.

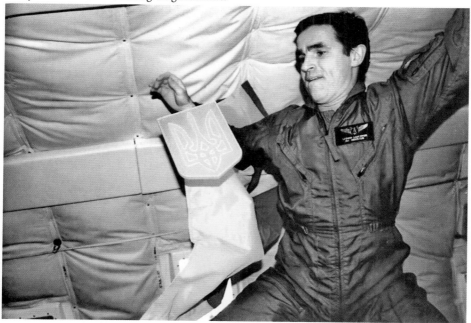

列昂尼德·卡登纽克（Leonid Kadenyuk）是乌克兰独立后的首位宇航员。1997年，作为国际任务STS-87的一部分，他乘坐美国宇航局的"哥伦比亚"号航天飞机完成了其首次飞行。图为在失重状态下的他和乌克兰标志。

PROJECTILE MOTION
抛物运动

TERMS 术语

projectile	抛物体
path	路径
curved path	曲线路径
parabolic path	抛物线路径
trajectory	轨迹
parabola	抛物线
horizontal direction	水平方向
horizontal component of velocity	速度的水平分量
vertical component of velocity	速度的垂直分量
horizontal position	水平位置
vertical position	垂直位置
horizontal range	水平距离
maximum height	最大高度
peak of trajectory	轨迹最高点

LANGUAGE OF PHYSICS 物理学用语

Projectile motion is the motion of a body thrown or fired with an initial velocity in a uniform gravitational field.

抛物运动：物体以某一初速度在均匀重力场中被抛掷或发射的运动。

Projectile motion is the superposition of two motions: constant velocity motion in the horizontal direction and the motion of an object freely falling in the vertical direction under constant free-fall acceleration.

抛物运动是两种运动的叠加，即水平方向做匀速运动，垂直方向以恒定自由落体加速度做自由落体运动。

A **trajectory** is the path through space followed by a projectile.

轨迹：抛物体在空间中运动的路径。

The **trajectory of a projectile** in a uniform gravitational field is always a section of a parabola.

在均匀重力场中，抛物体轨迹总是抛物线的一段。

The **range** of a projectile is the maximum distance that a projectile will move in a horizontal direction.

抛物体的射程是抛物体在水平方向上运动的最大距离。

PROJECTILE TRAJECTORY

The graph shows the trajectory of a projectile fired from the origin at $t=0$ with an initial velocity v_0. The maximum height of the projectile is h, and the horizontal range is R. At the peak of the trajectory, the particle has coordinates $(R/2, h)$.

抛物体轨迹

该图显示了抛物体在$t=0$时刻从原点以初速度v_0发射的轨迹。抛物体的最大高度为h，水平距离为R。在轨迹最高点，质点的坐标为$(R/2, h)$。

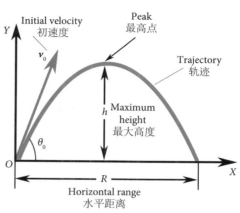

PROJECTILE VELOCITY

Projectile motion may be analyzed in terms of the horizontal and vertical components of velocity.

抛物体速度

抛物运动可以用速度的水平分量和垂直分量来分析。

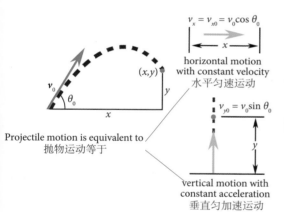

An example of projectile motion in real life. 现实生活中抛物运动的例子。

UNIFORM CIRCULAR MOTION
匀速圆周运动

TERMS	术语
circular path	圆周路径
curved path	曲线路径
centripetal acceleration	向心加速度
radial acceleration	径向加速度
tangential acceleration	切向加速度
total acceleration	总加速度
centripetal force	向心力
change in speed	速率变化
change in direction of velocity vector	速度矢量方向变化
center of curvature	曲率中心
center of rotation	旋转中心
radius of curvature	曲率半径
tangent to path	路径的切线

LANGUAGE OF PHYSICS 物理学用语

Uniform circular motion is motion in a circle at constant speed.

匀速圆周运动：在圆周上的匀速率运动。

Centripetal (radial) acceleration is the acceleration of a body moving with uniform circular motion.

向心(径向)加速度：物体做匀速圆周运动的加速度。

The **radial acceleration** is due to the change in the direction of the velocity vector and its direction is always toward the center of the circle.

径向加速度是由于速度矢量方向发生改变而产生的，它的方向总是指向圆心。

The **tangential acceleration** arises from the change in the speed of the particle and its direction is either in the same direction as the velocity vector (if the speed is increasing) or opposite (if the speed is decreasing).

切向加速度源自质点速率的变化，其方向与速度矢量相同（如果速率增大）或相反（如果速率减小）。

A **centripetal force** is necessary to cause an object to move in a circle at constant speed.

向心力是使一个物体做匀速率圆周运动所必需的。

The **centrifugal force** is a reaction force arising from the centripetal force.

离心力：由向心力引起的反作用力。

力学 43

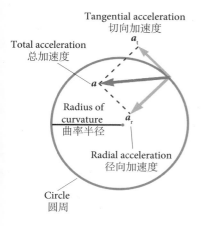

CIRCULAR MOTION

The total acceleration consists of a radial component arising from the change in the direction of velocity and directed toward the center of rotation, and a tangential component arising from the change in the speed of the particle. If the speed of motion changes, then so will the radius of the circle.

圆周运动

总加速度包括由于速度方向的改变而产生的、朝向旋转中心的径向分量和由于质点速率的改变而产生的切向分量。如果运动的速率改变，那么圆周的半径也会改变。

Neglecting forces arising from gravity, uniform circular motion can be seen in a Ferris wheel.

忽略重力作用，可将摩天轮视作做匀速圆周运动。

CURVED MOTION

As the particle moves along a curved path, the acceleration vector can be resolved into two component vectors: a radial component vector and a tangential component vector.

曲线运动

当质点沿曲线运动时，加速度矢量可分解为两个分矢量：一个径向分矢量和一个切向分矢量。

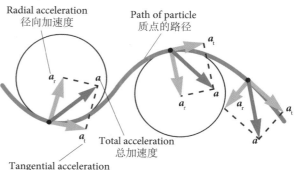

FORCE
力

TERMS	术语
force	力
external force	外力
internal force	内力
contact force	接触力
field force	场力
fundamental force	基本力
gravitational force	重力
electromagnetic force	电磁力
strong nuclear force	强核力
weak nuclear force	弱核力
friction force	摩擦力
static friction force	静摩擦力
kinetic friction force	动摩擦力
normal force	法向力
tension force	张力
resistive force, drag	阻力
central force	有心力
fictitious force	假力
centripetal force	向心力
centrifugal force	离心力
conservative force	保守力
nonconservative force	非保守力
restoring force	回复力
resultant force	合力
net force	净力

LANGUAGE OF PHYSICS
物理学用语

A **force** is any interaction that, when unopposed, will change the motion of an object. A force can cause an object with mass to change its velocity, i.e. to accelerate.

力：任意一种相互作用，无反作用时能改变物体的运动状态。力可使有质量的物体改变速度，即产生加速度。

A **force** has both magnitude and direction, making it a vector quantity.

力既有大小，又有方向，因此力是一个矢量。

The only known **fundamental forces** in nature are all field, i.e. non-contact forces: gravitational attraction between masses; electromagnetic forces between electric charges; strong nuclear forces between subatomic particles; weak nuclear forces that are manifested in radioactive decay processes.

自然界已知的**基本力**都是场力，即非接触力：有质量物体之间的万有引力、电荷间的电磁力、亚原子粒子之间的强核力、放射性衰变过程中的弱核力。

EXAMPLES OF FORCES

Some examples of forces applied to various objects.

(a), (b), (c)
Forces, which do not involve physical contact between two objects, but act through empty space are known as field forces.

(d), (e), (f)
Another class of forces represents the result of physical contact between two objects and these are known as contact forces.

力的例子

应用于各种物体的力举例。

(a)、(b)、(c)
不涉及两物体之间的物理接触，而是通过真空作用的力，称为场力。

(d)、(e)、(f)
另一种力是两个物体通过物理接触而产生的，称为接触力。

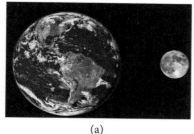

(a)

(d)

(b)

(e)

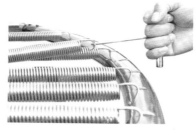

(c)

(f)

NEWTON'S FIRST LAW
牛顿第一定律

TERMS	术语
law of motion	运动定律
law of inertia	惯性定律
external force	外力
inertial frame	惯性参考系
inertia	惯性
body at rest	静止的物体
body in motion	运动的物体
inertial mass	惯性质量
gravitational mass	引力质量
weight	重量

LANGUAGE OF PHYSICS   物理学用语

Law of inertia
A body at rest remains at rest, and a body in motion will continue in motion with a constant velocity, unless it experiences a net external force. This is known as **Newton's First Law**.

惯性定律
物体所受外力为零时，物体维持运动状态不变，即保持静止或匀速直线运动。该定律又称为**牛顿第一定律**。

An **external force** is a force exerted on the object by another object.

外力：一个物体被另一个物体施加的力。

An **inertial frame of reference** is one in which Newton's First Law is valid. This is an unaccelerated frame.

惯性参考系：牛顿第一定律能严格成立的参考系。该参考系无加速度。

Inertia is the resistance of any physical object to any change in its velocity.

惯性：物体抵抗其速度被改变的性质。

The inertia of an object is given by its **mass**.

物体的惯性由其**质量**来衡量。

The **weight** of a body is equal to the magnitude of the gravitational force exerted by the Earth on the body and varies with location.

物体的**重量**等于其受到的重力大小，且随其所处位置的变化而不同。

力学 **47**

An object at rest will remain at rest ...

静止的物体依旧静止……

... unless acted on by an unbalanced force.

……除非受到不为零的合外力作用。

An object in motion
will continue with constant speed and direction ...

运动的物体继续保持运动速率和方向不变……

... unless acted on by an unbalanced force.

……除非受到不为零的合外力作用。

NEWTON'S SECOND LAW
牛顿第二定律

TERMS	术语
acceleration	加速度
net force	净力
mass	质量
directly proportional	成正比的
inversely proportional	成反比的
newton	牛顿
dyne	达因
free-body diagram	自由物体受力图
frictionless pulley	无摩擦的滑轮
inclined plane	斜面
massless string	无质量的轻绳

LANGUAGE OF PHYSICS 物理学用语

Newton's Second Law
The acceleration of an object is directly proportional to the net force acting on it and inversely proportional to its mass.

牛顿第二定律
物体加速度的大小与其受到的净力成正比，与其质量成反比。

The more general form of Newton's Second Law
The force is equal to the rate of change of momentum of an object. This is more applicable in situations where an object does not have a fixed mass e.g. a rocket.

牛顿第二定律更广义的表述
力等于物体动量的变化率。该表述更适用于质量不固定的物体，如火箭。

A **free-body diagram** is a graphical method for solving problems involving forces. The most important step is to draw only those forces that act on the object that you are isolating.

自由物体受力图：解决受力问题的作图法。其最重要的步骤是只画出被隔离物体的受力情况。

MOST IMPORTANT LAW OF CLASSICAL MECHANICS

Newton's Second Law answers the question of what happens to an object that has a nonzero resultant force acting on it.

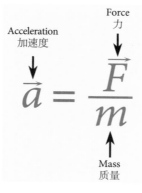

经典力学最重要的定律

牛顿第二定律回答了合力不为零时物体如何运动的问题。

FREE-BODY DIAGRAM

(a) Two masses connected by a massless string over a frictionless pulley.

(b) Free-body diagram for m_1.

(c) Free-body diagram for m_2.

自由物体受力图

(a) 两个物体由一根无质量的轻绳连接，跨过一个无摩擦的滑轮。

(b) m_1 的自由物体受力图。

(c) m_2 的自由物体受力图。

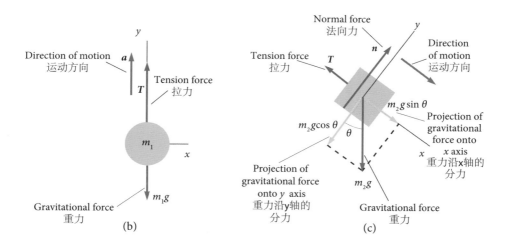

NEWTON'S THIRD LAW
牛顿第三定律

TERMS	术语
action force	作用力
reaction force	反作用力
law of action and reaction	作用力和反作用力定律
pairwise interaction	两两相互作用
normal force	法向力
thrust	推力
rocket propulsion	火箭推进

LANGUAGE OF PHYSICS
物理学用语

Newton's Third Law
If two bodies interact, the force exerted on object 1 by object 2 is equal and opposite to the force exerted on object 2 by object 1:

$$F_{12} = -F_{21}$$

Forces always occur in pairs.
A single isolated force cannot exist.

The action force is equal in magnitude to the reaction force and opposite in direction. In all cases, the action and reaction forces act on different objects.

牛顿第三定律
相互作用的两个物体，物体 1 和物体 2 所受到的力大小相等，方向相反：

$$F_{12} = -F_{21}$$

力总是成对出现。
不存在单个孤立的力。

作用力与反作用力大小相等、方向相反。在任何情况下，作用力和反作用力都作用在不同物体上。

力学 **51**

NEWTON'S THIRD LAW OF MOTION

The force exerted by object 1 on object 2 is equal to and opposite the force exerted by object 2 on object 1.

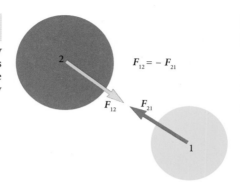

牛顿第三运动定律

物体1作用在物体2上的力和物体2作用在物体1上的力大小相等、方向相反。

ACTION-REACTION PAIR

The action force is equal in magnitude to the reaction force and opposite in direction.

作用与反作用

作用力和反作用力大小相等、方向相反。

The thrust on the rocket is the force exerted on the rocket by the ejected exhaust gases. When a rocket moves in a vacuum, it loses some mass due to the ejected gases. The ejected gases acquire some momentum, so the rocket receives a balancing momentum in the opposite direction. Therefore, the thrust from the exhaust gases results in the acceleration of the rocket.

火箭受到的推力是其喷出的燃料燃烧产生的气体对火箭的作用力。在真空中运动时，火箭由于喷射气体会损失一些质量。喷射出的气体获得一些动量，而火箭受到一个反向的动量与之平衡。因此，喷射气体产生的推力使火箭加速。

WORK
功

TERMS	术语
work done by a force	力做的功
work done by a constant force	恒力做功
work done by a varying force	变力做功
total work	总功
net work	净功
work done by a spring	弹簧做功
force acting on an object	作用于物体的力
displacement	位移
direction of displacement	位移方向
area under a curve	曲线下的面积

LANGUAGE OF PHYSICS
物理学用语

The **work done by a constant force** is defined as the product of this force acting on a body in the direction of a displacement, times the displacement of the body.

恒力做工：物体在位移方向上受到的力与位移的乘积。

The **work done by a varying force** acting on a body moving along the x direction from initial to final position is the line integral of the component of this force along the x direction.

变力做功：变力使物体沿x方向由初始位置运动到最终位置做的功，等于该力沿x方向的分量的线积分。

The **net work** done by several forces acting on a body is the sum of the individual amounts of work done by each force.

净功：几个力共同作用在物体上做的功等于每个力单独做功的叠加。

Work is an energy transfer.

功：一种能量转换。

Work is positive if energy is transferred to the system.

正功：使系统能量增加。

Work is negative if energy is transferred from the system.

负功：使系统能量减少。

Work is zero when the force is perpendicular to the displacement or the body does not move.

当外力垂直于物体的位移或者物体位置不变时，外力做**功**为零。

Work is a scalar quantity.

功是标量。

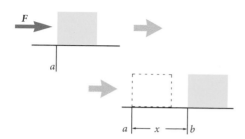

WORK DONE BY A CONSTANT FORCE

If a force acting on an object undergoes a displacement x, the work done by the force F is $F \cdot x$.

恒力做功

如果力F使一个物体产生位移x，则该力做的功为$F \cdot x$。

WORK DONE BY A VARYING FORCE

(a) The work done by the force F for the small displacement Δx equals the area of the shaded rectangle.
(b) The work done by the variable force F as the particle moves from x_i to x_f is exactly equal to the area under this curve.

变力做功

(a) 力F在小段位移Δx上做的功等于阴影矩形的面积。
(b) 变力F使质点从x_i运动到x_f所做的功等于曲线下的面积。

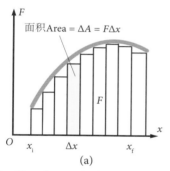

(a)

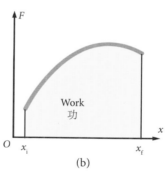

(b)

Oleg Skavish, a multiple Ukrainian record-holder, moved himself from his original position and dragged a tram of 38 tonnes along the way.

奥列格（Oleg Skavish），多项乌克兰纪录保持者，拉着一辆38吨重的电车，从初始位置移动。

ENERGY
能

TERMS	术语
energy	能
kinetic energy	动能
potential energy	势能
gravitational potential energy	重力势能
elastic potential energy	弹性势能
total mechanical energy	总机械能
relativistic kinetic energy	相对论动能
change in energy	能量变化
work-energy theorem	功能定理
power	功率
average power	平均功率
instantaneous power	瞬时功率

LANGUAGE OF PHYSICS 物理学用语

Energy is the ability of a body or system of bodies to perform work.

能：物体或物体所在系统做功的本领。

Kinetic energy is the energy that a body possesses by virtue of its motion. The kinetic energy is equal to the work that must be done to bring the body from rest into that state of motion.

动能：物体由于运动而拥有的能量。动能等于使物体从静止状态进入运动状态所必须做的功。

Gravitational potential energy is the energy that an object has due to its position in a gravitational field. The potential energy is equal to the work that must be done to bring the body from infinity into that particular position.

重力势能：物体由于其在重力场中的位置而产生的能量。势能等于把物体从无穷远处移到特定位置所做的功。

Elastic potential energy is energy stored in a spring by virtue of its stiffness.

弹性势能：由于弹簧的劲度而储存在弹簧中的能量。

The **work-energy theorem** states that the net work done on a particle by external forces equals the change in the kinetic energy of the particle.

功能定理表示，外力对质点所做的净功等于该质点动能的变化量。

Power is the time rate of doing work.
Power is the time rate of energy transfer.

功率：做功随时间的变化率，是能量转换的速率。

KINETIC ENERGY

Kinetic energy is the energy associated with the motion of a body.

动能

动能是与物体的运动有关的能量。

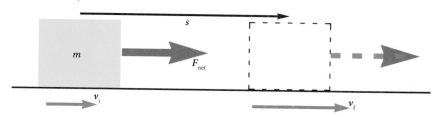

The hammer has kinetic energy associated with its motion and can do work on the nail, driving it into the floor.

锤子因为运动而拥有动能,可以对钉子做功,把钉子钉进地板。

POTENTIAL ENERGY

Potential energy is the energy associated with the position or configuration of an object.

势能

势能是与物体的位置或形状有关的能量。

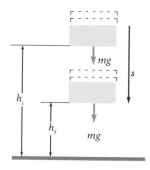

Manjavskyj falls in the Carpathian Mountains. The gravitational potential energy of the water at the top of the falls is converted to kinetic energy at the bottom, and then into thermal energy and a little sound energy.

曼贾夫斯基瀑布(Manjavskyj falls)位于喀尔巴阡山脉。

瀑布顶部水的重力势能转化为底部的动能,然后进一步转化为热能和少量声能。

CONSERVATION of ENERGY
能量守恒

TERMS 术语

transformation of energy	能量转换
conservation of energy	能量守恒
loss of energy	能量损耗
isolated system	孤立系统
conservative force	保守力
nonconservative force	非保守力
loss of kinetic energy	动能损耗

LANGUAGE OF PHYSICS 物理学用语

A **system** is an aggregate of two or more particles that is treated as an individual unit.

系统：两个及两个以上被视为独立单元的质点的集合。

A **closed system** is an isolated system that is not affected by any external influences.

封闭系统：一个不受外界影响的孤立系统。

A **force** is **conservative** if the work it does on a particle is independent of the path the particle takes between two points.

保守力：对一个质点做的功与两点间路径无关的力。

The **total mechanical energy** of a system is the sum of the kinetic energy and potential energy.

总机械能：系统内动能和势能的总和。

Law of conservation of energy
In any closed system, the total energy of the system remains constant.
or
If no external forces do work on a system, and there are no nonconservative forces, the total mechanical energy of the system is constant.

能量守恒定律
在任何封闭系统内，系统的总能量保持不变。
或
一个系统没有外力对其做功，也没有非保守力，那么该系统的总机械能保持不变。

The **change in the total mechanical energy** of a system equals the change in the kinetic energy due to internal nonconservative forces plus the change in kinetic energy due to all external forces.

系统总机械能的变化量等于系统内非保守力和所有外力造成的动能变化之和。

TRANSFORMATION OF ENERGY

When a ball falls, its total energy (the sum of the kinetic and potential energies) remains constant and equal to its initial potential energy.

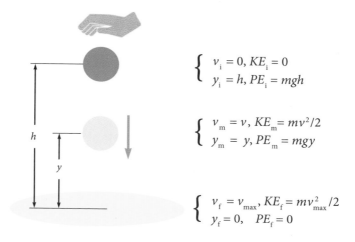

$$\begin{cases} v_i = 0, KE_i = 0 \\ y_i = h, PE_i = mgh \end{cases}$$

$$\begin{cases} v_m = v, KE_m = mv^2/2 \\ y_m = y, PE_m = mgy \end{cases}$$

$$\begin{cases} v_f = v_{max}, KE_f = mv_{max}^2/2 \\ y_f = 0, \quad PE_f = 0 \end{cases}$$

能量转换

当小球掉落时,总能量(动能与势能之和)保持不变且等于其初始势能。

Multiexposure photograph of a diver. During this process, several types of energy transformations occur. Gravitational potential energy is associated with the change in vertical position of the diver relative to the Earth. Elastic potential energy is evident in the bending of the springboard.

跳水运动员的多重曝光照片。跳水过程中发生了多种能量转换。重力势能与跳水运动员相对于地球的垂直距离的变化有关。弹性势能从跳板的弯曲中显而易见。

LINEAR MOMENTUM
线性动量

TERMS	术语
momentum	动量
linear momentum	线性动量
total momentum	总动量
initial momentum	初动量
final momentum	末动量
isolated system	孤立系统
internal force	内力
external force	外力
law of conservation of momentum	动量守恒定律
impulse of a force	力的冲量
impulse-momentum theorem	冲量-动量定理

LANGUAGE OF PHYSICS　　　　　　　　　　物理学用语

The **linear momentum** of a particle of mass m moving with a velocity v is defined to be the product of the mass and velocity. It is a vector quantity.

线性动量：一个质量为m的质点以速度v运动，其线性动量等于其质量和速度的乘积。它是一个矢量。

Conservation of momentum
Whenever two isolated, uncharged particles interact with each other, their total momentum remains constant.
or
The total momentum of an isolated system is constant.

动量守恒
在任何时候，两个孤立且不带电的质点相互作用，它们的总动量保持不变。
或
一个孤立系统的总动量保持不变。

Impulse is the product of the force that is acting and the time that the force is acting.

冲量：作用力与其作用时间的乘积。

Impulse-momentum theorem
The impulse of the force equals the change in the momentum of the particle.

冲量-动量定理
质点所受力的冲量等于其动量的变化量。

Internal forces are forces that originate within the system and act on the particles within the system.

内力：源于系统内并作用于系统内质点的力。

External forces are forces that originate outside the system and act on the system.

外力：源于系统外并作用于系统上的力。

力学 **59**

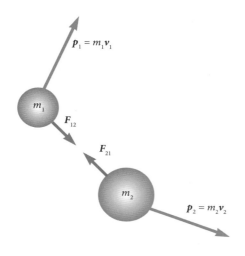

CONSERVATION OF MOMENTUM

At some instant, the momentum of m_1 is $p_1 = m_1 v_1$ and the momentum of m_2 is $p_2 = m_2 v_2$. The particles exert a force on each other according to Newton's Third Law $F_{12} = -F_{21}$. No external forces are present. Since the forces are equal and opposite and act for the same time, the total momentum of the system remains unchanged.

动量守恒

例如，m_1的动量是$p_1 = m_1 v_1$，m_2的动量是$p_2 = m_2 v_2$。根据牛顿第三定律，两个质点间的相互作用力$F_{12} = -F_{21}$。当无外力存在时，由于两个力大小相等，方向相反，且作用时间相同，所以系统的总动量保持不变。

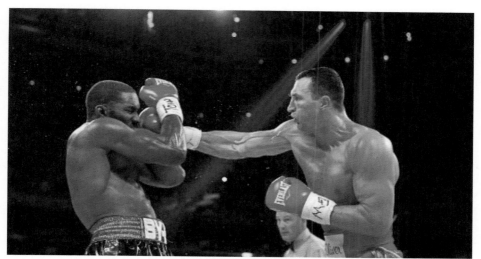

A boxer wisely moves his head backward just before receiving a punch from the Ukrainian boxing champion Vladimir Klitschko. As the impulse-momentum theorem tells us the average force exerted on the boxer's head is reduced when he extends the time of contact by moving away from the opponent's fist, assuming that the total momentum change is the same.

一位拳击手明智地将头后移，以免遭受乌克兰拳击冠军弗拉基米尔·克里钦科（Vladimir Klitschko）的一拳重击。正如冲量-动量定理告诉我们的，假设动量的变化量不变，在他向远离对手拳头的方向移动时，外力在他头上的作用时间延长了，所以平均作用力也就减小了。

COLLISION
碰撞

TERMS	术语
collision	碰撞
inelastic collision	非弹性碰撞
perfectly inelastic collision	完全非弹性碰撞
elastic collision	弹性碰撞
perfectly elastic collision	完全弹性碰撞
head-on collision	正碰撞
two-dimensional collision	二维碰撞
glancing collision	擦边碰撞
colliding particle	碰撞粒子
physical contact	物理接触
change in momentum	动量变化量
total momentum	总动量

LANGUAGE OF PHYSICS
物理学用语

A **collision** is when two particles come together for a short time and thereby produce impulsive forces on each other.

碰撞：两个粒子短时间接触，从而产生的相互的撞击力。

An **impulsive force** is much stronger than any other force available at that point in time and acts on an object for a very short interval during the collision.

撞击力在物体碰撞时作用于碰撞点上，时间短且强度远超该物体上的其他任何作用力。

An **inelastic collision** is one in which mechanical energy is not constant and some kinetic energy is lost. The velocity of separation of the two bodies is not equal to the velocity of approach.

非弹性碰撞：机械能不守恒，动能有损失。两个物体分离时的速度不等于接近时的速度。

A **perfectly inelastic collision** corresponds to the situation where the colliding bodies stick together after the collision. As much kinetic energy is lost as is theoretically possible during such a collision.

完全非弹性碰撞：碰撞后物体粘在一起的情况。此碰撞损失了理论上可能的所有动能。

An **elastic collision** is one in which kinetic energy is constant. The velocity of separation of the two bodies is equal to the velocity of approach. Momentum is conserved in all collisions for which there are no external forces.

弹性碰撞时动能守恒。两个物体分离时的速度等于接近时的速度。在无外力作用时，所有碰撞过程的动量均守恒。

ELASTIC COLLISION

Schematic representation of an elastic head-on collision between two particles.

弹性碰撞

两个粒子弹性正碰撞示意图。

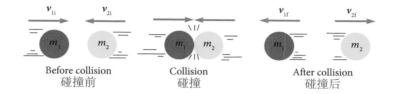

| Before collision | Collision | After collision |
| 碰撞前 | 碰撞 | 碰撞后 |

INELASTIC COLLISION

Schematic representation of a perfectly inelastic head-on collision between two particles.

非弹性碰撞

两个粒子完全非弹性正碰撞示意图。

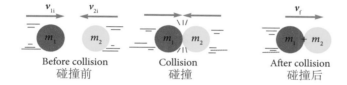

| Before collision | Collision | After collision |
| 碰撞前 | 碰撞 | 碰撞后 |

A section of the Large Hadron Collider. Colliders are used as research tools in particle physics by accelerating particles to very high kinetic energy and letting them impact other particles. Analysis of the byproducts of these collisions gives scientists good evidence of the structure of the subatomic world and the laws of nature governing it.

大型强子对撞机的部分结构。对撞机作为研究粒子物理学的工具，可以将粒子加速，使其具有非常高的动能，然后撞击其他粒子。对碰撞产物的分析为科学家认识亚原子世界的结构及其内在规律提供了确凿的证据。

CENTER of MASS
质心

TERMS	术语
system of particles	质点系
total mass of the system	系统的总质量
center of mass of the system	系统的质心
position of the center of mass	质心的位置
coordinates of the center of mass	质心的坐标
vector position of the center of mass	质心的矢量位置
velocity of the center of mass	质心的速度
acceleration of the center of mass	质心的加速度
total momentum of a system of particles	质点系的总动量
Newton's Second Law for a system of particles	质点系的牛顿第二定律
symmetric object	对称物体
axis of symmetry	对称轴
plane of symmetry	对称面
center of gravity	重心
imaginary particle	假想粒子

LANGUAGE OF PHYSICS   物理学用语

The **center of mass** is the point of a body at which all the mass of the body appears to be concentrated.

质心：物体上的一个点，假设该物体的所有质量都集中于这一点上。

The **center of mass of any symmetric object** lies on the intersection of an axis of symmetry and any plane of symmetry.

对称物体的质心位于对称轴与任意对称面的交点上。

The **center of mass moves** like an imaginary particle of mass equal to the total mass of the system under the influence of the resultant external force on the system.

质心运动起来视作假想粒子，其质量等于在合外力作用下整个系统的质量之和。

The **resultant force on a system of particles** equals the total mass of the system multiplied by the acceleration of the center of mass.

一个**质点系所受的合力**等于该系统的总质量乘以质心加速度。

The **total linear momentum of the system** equals the total mass multiplied by the velocity of the center of mass.

系统总的线性动量等于总质量乘以质心速度。

The **total momentum of the system** is constant if there are no external forces acting on it.

没有外力作用时，**系统总动量**保持不变。

CENTER OF MASS

The acrobat's center of mass follows a parabolic path, the same path that a particle would follow.

Parabolic path
抛物线路径

质心

杂技演员的质心以抛物线路径运动,质点遵循同样的运动路径。

CENTER OF MASS DETERMINATION

An experimental technique for determining the center of mass of a wrench. The wrench is hung freely from two different pivots, A and C. The intersection of the two vertical lines AB and CD locates the center of mass.

质心的确定

确定扳手质心的实验技巧。将扳手自由悬挂在两个不同支点 A 和 C 上。两条垂线 AB 和 CD 的交点就是质心。

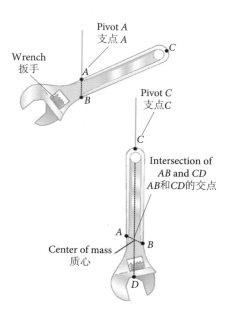

Pivot A
支点 A

Wrench
扳手

Pivot C
支点 C

Intersection of AB and CD
AB 和 CD 的交点

Center of mass
质心

STATIC EQUILIBRIUM
静平衡

TERMS 术语

object in equilibrium	处于平衡状态的物体
condition of equilibrium	平衡状态
condition of translational equilibrium	平动平衡的条件
condition of rotational equilibrium	转动平衡的条件
point of application of a force	力的作用点
line of action of a force	力的作用线
coplanar force	共面力
equivalent force	等价力
concurrent force	共点力
couple	力偶
resultant external force	合外力
resultant external torque	合外力矩
unstable (stable) equilibria	不稳定（稳定）平衡
neutral equilibria	随遇平衡

LANGUAGE OF PHYSICS 物理学用语

The term **equilibrium** means either that the object is at rest or that its center of mass moves with constant velocity.

平衡：物体静止，或者物体质心以恒定的速度运动。

A **body is in equilibrium** under the action of several forces if the body has zero translational acceleration and no rotational acceleration.

平衡物体：如果在几个力的作用下，物体既没有平动加速度也没有转动加速度，则称该物体处于平衡状态。

Condition of translational equilibrium
If the resultant external force on an object is zero.

平动平衡的条件
物体所受的合外力为零。

Condition of rotational equilibrium
If the resultant external torque on an object is zero about any origin.

转动平衡的条件
围绕物体上的任意一点，其所受的合外力矩为零。

If **two forces act on an object**, the object is in equilibrium if and only if the forces are equal in magnitude, opposite in direction, and have the same line of action.

对于**作用在同一物体上的两个力**，当且仅当它们大小相等、方向相反且作用在同一条直线上时，才能使该物体处于平衡状态。

If **an object subjected to three forces is in equilibrium**, those forces must be concurrent.

如果**一个物体同时受到三个力的作用仍处于平衡状态**，则这些力必然作用于同一点上。

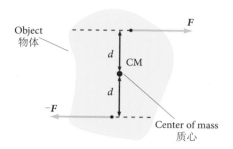

COUPLE

Two equal and opposite forces acting on the object form a couple. In this figure the object rotates clockwise.

力偶

作用在同一物体上的两个大小相等、方向相反的力构成一对力偶。如图所示，该物体将顺时针旋转。

EQUILIBRIUM

(a) The object is not in equilibrium because the two forces do not have the same line of action.
(b) The object is in equilibrium because the two forces act along the same line of action.

平衡

(a) 该物体处于不平衡状态是因为两个力没有作用在同一条直线上。
(b) 该物体处于平衡状态是因为两个力作用在同一条直线上。

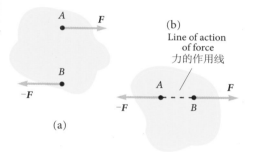

This large balanced rock at the Dovbush Rocks in the Carpathian Mountains, an example of stable equilibrium.

这块位于喀尔巴阡山脉的多福氏山上的巨大岩石，正是稳定平衡的例子。

ROTATION of a RIGID OBJECT
刚体的转动

TERMS	术语
rigid object	刚体
rigid-object model	刚体模型
nondeformable object	不发生形变的物体
pure rotational motion	纯转动
axis of rotation	旋转轴
fixed axis	固定轴
rotational kinematics equation	转动的运动学方程
translation of center of mass	质心的平动
rotation about center of mass	围绕质心的转动
circular motion	圆周运动
circular arc	圆弧

LANGUAGE OF PHYSICS   物理学用语

A **rigid object** is one that is nondeformable or in which the separation between all pairs of particles remains constant.

刚体：不发生形变，或内部各点的相对位置保持不变的物体。

Rotation about a fixed axis means that all the particles of the body, except those which lie on the axis of rotation, move along circular paths.

定轴转动：物体上的所有点都做圆周运动，位于旋转轴上的点除外。

The **axis of rotation** is the straight line through all fixed points of a rotating rigid body around which all other points of the body move in circles.

旋转轴：穿过旋转刚体的所有固定点的直线，该刚体的所有其他点都绕它做圆周运动。

The **most general motion of a rigid object** is composed of the translation of the center of mass of the object plus its rotation about the center of mass.

刚体最常见的运动包括物体质心的平动和围绕质心的转动。

When an extended object rotates about its axis at any given time, different parts of the object have different velocities and accelerations.

扩展物体在任意给定时间内绕其轴旋转时，物体的不同部位具有不同的速度和加速度。

ROTATIONAL MOTION

Rotation of a rigid object about a fixed axis. A particle at B rotates in a circle of radius R centered at O. Every particle on the object undergoes circular motion about O.

转动

刚体的定轴转动。若刚体上的B点绕O点做半径为R的圆周运动,则刚体上的每个点都围绕O点做圆周运动。

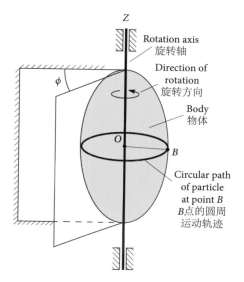

The Ukrainian skating pair Julia Lavrentjeva and Yuriy Rudyk. In this circular motion the female skater is spun in a circular arc by her partner with her body nearly parallel to the ice.

乌克兰滑冰组合朱莉娅·拉夫伦蒂耶娃（Julia Lavrentjeva）和尤里·鲁迪克（Yuriy Rudyk）。在这个圆周运动中,女运动员在其搭档的帮助下沿着圆弧旋转,她的身体几乎与冰面平行。

ANGULAR VELOCITY and ACCELERATION
角速度和角加速度

TERMS 术语	
angular position	角位置
angular displacement	角位移
angular speed	角速度
angular acceleration	角加速度
average angular acceleration	平均角加速度
instantaneous angular acceleration	瞬时角加速度
reference line	参考线
arc length	弧长
radius of circle	圆的半径
radius of arc	圆弧的半径
circumference	圆周
revolution	旋转
angle in degrees	以度为单位的角度
angle in radians	以弧度为单位的角度
clockwise rotation	顺时针旋转
counterclockwise rotation	逆时针旋转

LANGUAGE OF PHYSICS   物理学用语

Angular displacement is the angle that a body rotates through while in rotational motion.

角位移：物体在旋转运动中旋转的角度。

Angular velocity is the change in the angular displacement of a rotating body about the axis of rotation with time.

角速度：旋转物体绕旋转轴的角位移随时间的变化量。

Angular acceleration is the change in the angular velocity of a rotating body with time.

角加速度：旋转物体的角速度随时间的变化量。

When a rigid object rotates about a fixed axis, every part of the object has the **same angular speed and the same angular acceleration**. The directions of angular velocity and angular acceleration are along the axis of rotation.

做定轴转动时，刚体的每个部分都有**相同的角速度和角加速度**，且角速度和角加速度的方向沿着该旋转轴。

The **linear speed of a point on a rotating rigid object** equals the distance of that point from the axis of rotation multiplied by the angular speed.

旋转着的刚体上一点的线速度等于这个点到旋转轴的距离乘以角速度。

ANGULAR DISPLACEMENT

The position of a point A on a rotating body is defined by the angular displacement θ from some reference line that connects that point to the axis of rotation.

角位移

旋转物体上点A的位置，由连接点A与旋转轴的某条参考线的角位移θ确定。

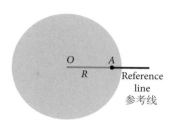

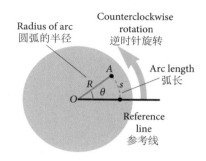

ANGULAR VELOCITY AND ANGULAR ACCELERATION

As a rigid object rotates about the fixed axis through O, the point B

角速度和角加速度

当一个刚体围绕通过O点的定轴转动时，点B

(a) has a linear velocity v that is always tangent to the circular path of radius R;

(b) experiences a tangential component of linear acceleration a_t and a centripetal component of linear acceleration a_r.

The total linear acceleration of this point is $a = a_t + a_r$.

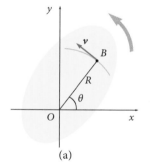

(a)

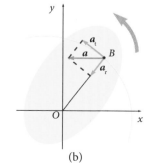

(b)

(a) 具有的线速v总是与半径为R的圆周路径相切；

(b) 其线加速度可分解为沿切向的a_t和沿径向的a_r。

该点总的线加速度$a = a_t + a_r$。

MOMENT of INERTIA
转动惯量

TERMS 术语

inertia	惯性
moment of inertia	转动惯量
moment of inertia of a system of particles	粒子系统的转动惯量
moment of inertia of a rigid object	刚体的转动惯量
volume density	体积密度
surface density	表面密度
linear density	线密度
parallel-axis theorem	平行轴定理
hoop	圆环
disk	圆盘
rectangular plate	矩形平板
long thin rod	细长杆
solid cylinder	实心圆柱体
hollow cylinder	空心圆柱体
cylindrical shell	圆柱壳
thin spherical shell	薄球壳
solid sphere	实心球
rotation axis	旋转轴
rotation axis through end	过端点的旋转轴
rotation axis through center	过中心的旋转轴
rotational energy	转动能

LANGUAGE OF PHYSICS   物理学用语

The **moment of inertia** is the measure of the resistance of a body to a change in its rotational motion.

转动惯量：测量物体在转动过程中对改变转动的力产生的阻力的物理量。

The **parallel-axis theorem:**
If the moment of inertia of a rigid body about an axis that passes through the center of mass is I_{CM}, then the moment of inertia I_Z about an axis that is parallel to and a distance D away from the axis that passes through the center of mass is

$$I_Z = I_{CM} + mD^2$$

平行轴定理：
如果一刚体绕通过其质心的旋转轴的转动惯量为I_{CM}，则其围绕与此轴距离为D且平行于此轴的新的旋转轴的转动惯量I_Z为

$$I_Z = I_{CM} + mD^2$$

The **kinetic energy of rotation** is the energy that a body possesses by virtue of its rotational motion.

旋转动能：物体由于转动而具有的能量。

MOMENTS OF INERTIA OF SOME RIGID OBJECTS   一些刚体的转动惯量

Long thin rod with rotation axis through center
围绕过中心的旋转轴转动的细长杆
$I_{CM} = \frac{1}{12}mL^2$

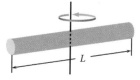

Solid cylinder or disk
实心圆柱体或圆盘
$I_{CM} = \frac{1}{2}mR^2$

Long thin rod with rotation axis through end
围绕过端点的旋转轴转动的细长杆
$I_{CM} = \frac{1}{3}mL^2$

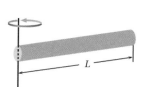

Hollow cylinder
空心圆柱体
$I_{CM} = \frac{1}{2}m(R_1^2 + R_2^2)$

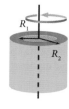

Thin spherical shell
薄球壳
$I_{CM} = \frac{2}{3}mR^2$

Hoop or cylindrical shell
圆环或圆柱壳
$I_{CM} = mR^2$

Solid sphere
实心球
$I_{CM} = \frac{2}{5}mR^2$

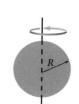

Rectangular plate
矩形平板
$I_{CM} = \frac{1}{12}m(a^2+b^2)$

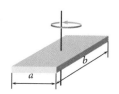

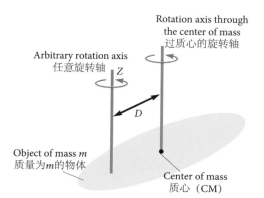

PARALLEL-AXIS THEOREM

Even for a highly symmetric object the calculations of moment of inertia about an arbitrary axis can be somewhat cumbersome. The parallel-axis theorem often simplifies these calculations.

平行轴定理

即使对于高度对称的物体，计算绕任意轴的转动惯量也有些烦琐。平行轴定理通常能简化计算过程。

TORQUE
力矩

TERMS	术语
torque (moment of force)	力矩
axis of rotation	旋转轴
reference axis	参考轴
pivot point	支点
point of application of a force	力的作用点
line of action of a force	力的作用线
moment arm	力（矩）臂
lever arm	杠杆臂
magnitude of torque	力矩大小
rotating tendency	旋转趋势
direction of force	力的方向

LANGUAGE OF PHYSICS 物理学用语

The **moment of force** acting on an object, often called **torque**, is the product of the force and the perpendicular distance between the line of action of the force from a reference point.

作用于物体的**力矩**等于力与参考点到力的作用线的垂直距离的乘积。

Torque is a measure of the tendency of a force to rotate the object about some axis.

力矩：衡量一个力使物体旋转的趋向的物理量。

Torque is the product of force and the moment arm of that force.
$$\tau = rF\sin\phi$$

力矩：力和力臂的乘积。
$$\tau = rF\sin\phi$$

Torque is the vector or cross product of position vector and force vector.

力矩：位置矢量与力矢量的矢量积或叉积。

Torque is defined only when a reference axis is specified.

只有先指定了参考轴才能定义**力矩**。

The **lever arm** is defined as the perpendicular distance from the axis of rotation to the direction or line of action of the force.

杠杆臂：旋转轴到力的作用方向或作用线的垂直距离。

The **line of action of a force** is an imaginary line extending out both ends of the vector representing the force.

力的作用线：一条假想的线，从代表力的矢量的两端向外延伸。

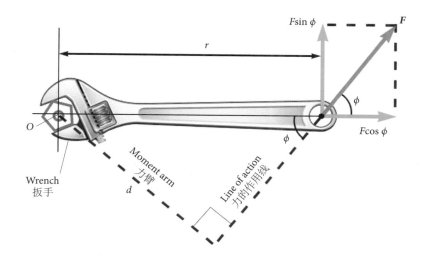

TORQUE

The force F has a greater rotating tendency about O as F increases and as the moment arm d increases. It is the component $F\sin\phi$ that tends to rotate the system about O.

力矩

随着力F的增大和力臂d的增大，力F围绕点O旋转的趋势增强。是作用力的分量$F\sin\phi$使系统绕点O旋转。

When a force is exerted on a rigid object pivoted about an axis, the object tends to rotate about that axis. The tendency of a force to rotate an object about some axis is measured by a torque.

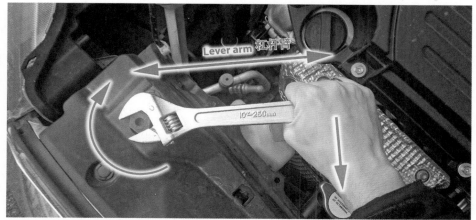

当力作用于绕轴旋转的刚体时，该物体倾向于绕此轴转动。使物体绕某个轴转动的力的大小通过力矩来衡量。

ROLLING MOTION
滚动

TERMS	术语
rolling motion	滚动
CM = center of mass	质心
rolling down	向下滚动
rolling on rough surface	在粗糙表面上滚动
rolling without slipping	无滑动的滚动
rolling body	滚动的物体
rolling friction	滚动摩擦
point on the rim	边缘上的点
cycloid path	摆线路径
rotation about CM	绕质心的转动
translation of CM	质心的平动
contact point	接触点
total kinetic energy	总动能
rotational kinetic energy	转动动能
translational kinetic energy	平动动能

LANGUAGE OF PHYSICS   物理学用语

A **rolling motion** is defined as a motion about an axis of rotation which is not fixed in space.

滚动：旋转轴在空间中不固定的运动。

A **rolling motion** can be described as a combination of rotation about the center of mass and translation of the center of mass.

滚动可以描述为绕质心的转动和质心平动的组合。

The **trajectory** of any point of the body undergoing rolling motion in a plane is a **trochoid**; in particular, the trajectory of the CM is a **straight line**, while the trajectory of any point on the object's rim is a **cycloid**.

在平面上滚动的物体上任一点的**轨迹**都是一条**次摆线**；特别是，质心的轨迹呈一条**直线**，而物体边缘上任意一点的轨迹都是一条**摆线**。

The **total kinetic energy of a rigid body that is rolling on a rough surface without slipping** equals the rotational kinetic energy about its center of mass plus the translational kinetic energy of the center of mass.

在粗糙表面上无滑动滚动的刚体的总动能等于绕其质心的转动动能加质心的平动动能。

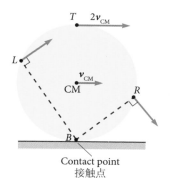

VELOCITY OF ROLLING BODY

The linear velocity of any point on a rolling body is in a direction perpendicular to the line from that point to the contact point. The center of the body moves with a velocity v_{CM}, while the point T moves with a velocity $2v_{CM}$.

滚动物体的速度

滚动物体上任意一点的线速度的方向都垂直于从该点到接触点的连线。物体的质心以速度 v_{CM} 移动，而 T 点的速度为 $2v_{CM}$。

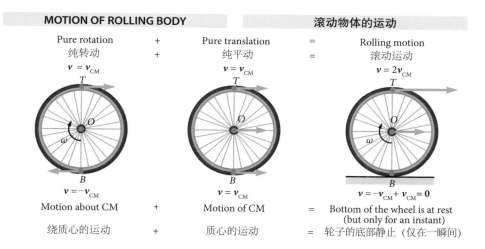

Light sources at the center and rim of a rolling ball illustrate the different paths these points take. The center moves in a straight line, as indicated by the green line, while a point on the rim moves in the path of a cycloid, as indicated by the red curve.

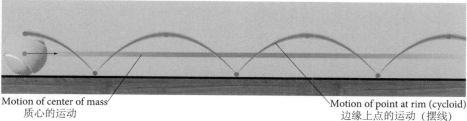

处于滚动的球的中心和边缘的不同光源说明这些点的运动路径不同。中心沿直线运动，如绿线所示，而边缘上的点沿摆线运动，如红色曲线所示。

ANGULAR MOMENTUM and its CONSERVATION
角动量和角动量守恒

TERMS	术语
momentum	动量
angular momentum	角动量
angular momentum of a particle	质点的角动量
angular momentum of a system of particles	质点系的角动量
isolated system	孤立系统
position vector about origin	相对于原点的位置矢量
net external torque	净外力矩
resultant external torque	合外力矩
time rate of change of angular momentum	角动量随时间的变化率
merry-go-round	旋转木马
gyroscope	陀螺仪
top	陀螺
conservation of angular momentum law	角动量守恒定律

LANGUAGE OF PHYSICS   物理学用语

Angular momentum is defined as the product of the moment of inertia of a rotating body and its angular velocity.

角动量：转动物体的转动惯量与其角速度的乘积。

Rotational analogue of Newton's Second Law for point particle
The resultant torque acting on a body about an axis through the center of mass equals the time rate of change of angular momentum regardless of the motion of the center of mass.

转动运动与牛顿第二定律的类比

作用在物体上的力对通过该物体质心的轴所产生的合力矩等于角动量随时间的变化率，与质心的运动无关。

Conservation of angular momentum law
The total angular momentum of a system is constant in both magnitude and direction if the resultant external torque acting on the system is zero, that is, if the system is isolated.

角动量守恒定律

如果作用于系统的合外力矩为零，即系统是孤立的，则系统的总角动量在大小和方向上都保持不变。

Constant angular momentum means that the final angular momentum is equal to the initial angular momentum.

恒定角动量意味着始末角动量相等。

MOTION OF ROLLING BODY

The angular momentum L of a particle of mass m relative to the origin O is defined by the cross product of the vector position of the particle r and its linear momentum p and is given by $L = r \times p$.

滚动物体的运动

质量为m的质点相对于原点O的角动量L等于质点的位置矢量r和其线性动量p的乘积，即$L = r \times p$。

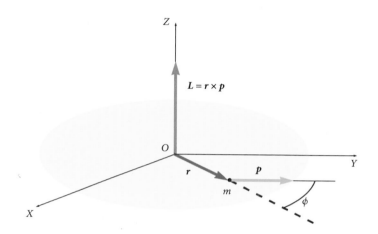

During this spinning action the ice skater's angular speed increases when she pulls her arms in close to her body, demonstrating that angular momentum is conserved.

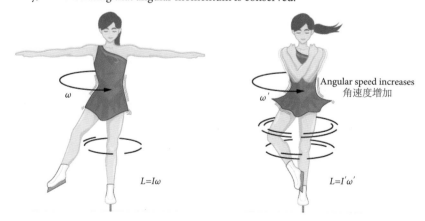

在自转过程中，滑冰运动员将手臂拉近身体时，角速度会增加，这表明角动量守恒。

NEWTON'S LAW of GRAVITY
牛顿万有引力定律

TERMS	术语
universal gravitational constant	万有引力常量
inverse-square law	平方反比（定）律
gravitational force	引力
conservative gravitational force	保守引力
central gravitational force	中心引力
gravitational field	引力场
gravitational wave	引力波
gravitational potential energy	引力位能
free-fall acceleration	自由落体加速度
attractive force	吸引力
weight	重量
black hole	黑洞
merging black holes	黑洞合并
theory of general relativity	广义相对论
energy transport	能量转移
gravitational radiation	引力辐射
electromagnetic radiation	电磁辐射

LANGUAGE OF PHYSICS   物理学用语

Newton's law of gravity
Every particle in the Universe attracts every other particle with a force that is directly proportional to the product of their masses and inversely proportional to the square of the distance between them.

牛顿万有引力定律
宇宙中任意两个质点之间存在吸引力，并且其大小与质点质量的乘积成正比，与质点间距离的平方成反比。

Free-fall acceleration decreases with increasing altitude. When the altitude approaches infinity the true **weight** approaches zero.

自由落体加速度：随着海拔高度增加而减小。当海拔趋近无穷时，物体的真实**重力**趋近于零。

A **geosynchronous satellite** is a satellite whose orbital motion is synchronized with the rotation of the Earth. This satellite is always over the same point on the equator as the Earth turns.

地球同步卫星的轨道运动与地球自转同步，并且总是在赤道上方同一点。

GRAVITATIONAL FORCE

The gravitational force is a field that always exists between two particles, regardless of the medium that separates them.

引力

在任意介质中，两个质点之间都存在引力。

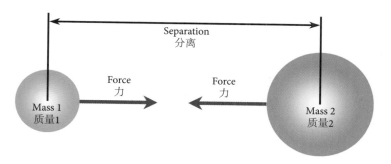

NEWTON'S LAW OF GRAVITY

The gravitational force decreases with the square of the separation and it is proportional to the mass of each particle.

牛顿万有引力定律

引力的大小与两个质点距离的平方成反比，与每个质点的质量成正比。

$$F = G \frac{m_1 m_2}{r^2}$$

On February 11, 2016, the LIGO Scientific Collaboration and Virgo Collaboration teams announced that they had made the first observation of gravitational waves, originating from a pair of merging black holes. Predicted in 1916 by Albert Einstein on the basis of his theory of general relativity, gravitational waves transport energy as gravitational radiation, a form of radiant energy similar to electromagnetic radiation.

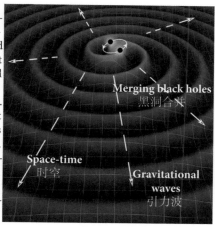

2016年2月11日，美国LIGO（激光干涉引力波天文台）科学合作组织和欧洲Virgo合作团队宣称他们首次探测到了引力波，称该引力波源自双黑洞合并。阿尔伯特·爱因斯坦（Albert Einstein）于1916年在广义相对论的基础上预言引力波会以引力辐射的形式（即一种类似于电磁辐射的辐射能形式）传播能量。

KEPLER'S LAWS
开普勒定律

TERMS 术语

motion of planets	行星运动
Sun	太阳
Moon	月球
orbit	轨道
circular orbit	圆形轨道
nearly circular orbit	近圆轨道
elliptical orbit	椭圆轨道
focal point	焦点
orbital period	轨道周期
semimajor axis	长半轴
semiminor axis	短半轴
orbital speed	轨道速度
escape speed	逃逸速度

LANGUAGE OF PHYSICS 物理学用语

Kepler's laws of planetary motion

1. All planets move in elliptical orbits with the Sun at one of the focal points.

2. The radius vector drawn from the Sun to a planet sweeps out equal areas in equal time intervals.

3. The square of the orbital period of any planet is proportional to the cube of the semimajor axis for an elliptical orbit.

The **escape velocity** is the minimum speed needed for an object to escape from the gravitational influence of a massive body. The escape velocity from the Earth is about 11.2 km/s.

A **black hole** is an object with an escape velocity greater than the speed of light.

开普勒行星运动定律

1. 所有行星都绕太阳做椭圆轨道运动，太阳位于椭圆的一个焦点上。

2. 从太阳到行星引一条半径矢量，则其在相等时间间隔内扫过的面积相等。

3. 所有行星轨道周期的平方与椭圆轨道的长半轴的立方成正比。

逃逸速度：物体脱离一个大质量物体的引力所需的最小速度。地球的逃逸速度大约为11.2 千米/秒。

黑洞：逃逸速度大于光速的物体。

KEPLER'S LAWS

Illustration of Kepler's three laws with planetary orbits.

开普勒定律

开普勒行星运动三大定律的图示。

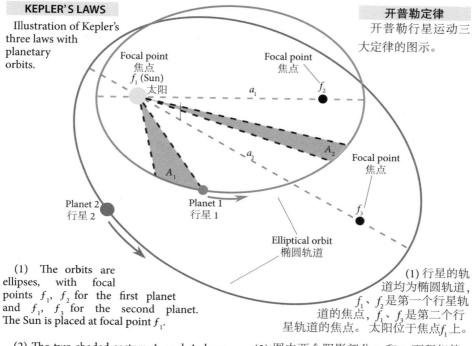

(1) The orbits are ellipses, with focal points f_1, f_2 for the first planet and f_1, f_3 for the second planet. The Sun is placed at focal point f_1.

(2) The two shaded sectors A_1 and A_2 have the same surface area. The time for planet 1 to cover segment A_1 is equal to the time to cover segment A_2.

(3) The total orbit times for planet 1 and planet 2 are in the ratio $a_1^{3/2} : a_2^{3/2}$.

(1) 行星的轨道均为椭圆轨道，f_1、f_2是第一个行星轨道的焦点，f_1、f_3是第二个行星轨道的焦点。太阳位于焦点f_1上。

(2) 图中两个阴影部分A_1和A_2面积相等。行星1扫过A_1部分的时间和扫过A_2部分的时间相等。

(3) 行星1与行星2的轨道周期之比为$a_1^{3/2} : a_2^{3/2}$。

John Houbolt explains Lunar orbit rendezvous (LOR). LOR is a key concept for landing humans on the Moon and returning them to the Earth and was first utilized for the Project Apollo missions in the 1960s and 1970s. LOR is first known to have been proposed in 1916 by Ukrainian rocket theoretician Yuri Kondratyuk, as the most economical way of landing humans on the Moon.

约翰·霍博尔特（John Houbolt）对月球轨道交会（Lunar orbit rendezvous，LOR）做了相关解释。月球轨道交会是人类登月及返航技术中的一个关键概念。20世纪六七十年代，这一概念在阿波罗计划中得到首次运用。月球轨道交会是人类登月最为经济的方式，最初是乌克兰火箭理论家尤里·孔德拉丘克（Yuri Kondratyuk）于1916年提出的。

ELASTIC PROPERTIES of SOLID
固体的弹性

TERMS 术语

deformation of solid	固体的形变
elasticity	弹性
compressibility	压缩性
stress	应力
strain	应变
cross-sectional area	横截面面积
constant of proportionality	比例系数
elastic modulus	弹性模量
ratio of stress to strain	应力应变比
elastic limit	弹性极限
ultimate breaking limit	断裂极限
tension	拉力
shear	剪切
compression	压缩

LANGUAGE OF PHYSICS 物理学用语

Elasticity is the property of a body by which it experiences a change in size or shape whenever a deforming force acts on it.

弹性：物体在外力的作用下大小或形状发生变化的一种特性。

Stress is a quantity proportional to the force producing a deformation. Stress is the force per unit cross-sectional area acting on an object.

应力：与导致形变的力成正比，是作用在物体单位横截面上的力。

Strain is the relative change in shape or size of an object due to externally-applied forces. Strain is a measure of the degree of deformation.

应变：在外力作用下物体的形状或大小的相对变化。应变是形变程度的量度。

Hooke's law
In an elastic body, the stress is directly proportional to the strain. Hooke's law is only valid below the elastic limit.

胡克定律
对于弹性物体，应力与应变成正比。胡克定律仅在弹性极限范围内成立。

The **elastic limit** is the maximum stress that can be applied to a substance before it becomes permanently deformed. When the stress exceeds the elastic limit, the material will not return to its original size or shape when the stress is removed.

弹性极限：材料在发生永久形变之前所能承受的最大应力。一旦应力超过了弹性极限，即使撤去应力，材料也无法恢复原来的大小或形状。

STRESS-STRAIN CURVE

The elastic limit is the point where the stress on a body becomes so great that the atoms of the body are pulled permanently away from their equilibrium position in the lattice structure. When the stress exceeds the elastic limit, the object is permanently distorted and the shape of the object is permanently changed. As the stress is increased even further, the material will ultimately break.

应力-应变曲线

弹性极限是使物体中的原子被永远拉离其在晶格中的平衡位置所需的应力。一旦应力超过弹性极限，物体就会产生永久形变。如果继续增加应力，材料最终将会断裂。

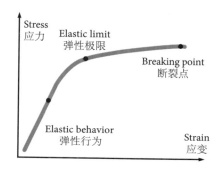

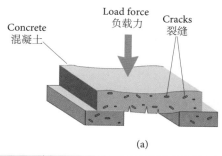

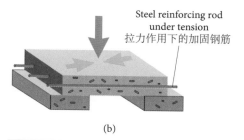

PRESTRESSED CONCRETE

(a) A concrete slab with no reinforcement tends to crack under a heavy load.
(b) The strength of the concrete is increased by using steel rods.

预应力混凝土

(a) 没有加固的混凝土板块在重荷下会产生裂缝。
(b) 使用钢筋后，混凝土的强度会提高。

The concrete is strengthened by prestressing it with steel rods under tension.

混凝土通过拉力下的钢筋预加应力，增大自身的强度。

ELASTIC MODULUS
弹性模量

TERMS 术语

Young's modulus	杨氏模量
shear modulus	剪切模量
bulk modulus	体积模量
elasticity in length	长度弹性
elasticity of shape	形状弹性
volume elasticity	体积弹性
stress	应力
tensile stress	拉应力
shear stress	切应力
volume stress	体积应力
strain	应变
tensile strain	拉应变
shear strain	切应变
volume strain	体积应变
angle of shear	剪切角
tangential force	切向力

LANGUAGE OF PHYSICS 物理学用语

Young's modulus measures the resistance of a solid to a change in its length.

杨氏模量：固体抵抗长度变化的能力的量度。

The **shear modulus** measures the resistance to motion of the layers in a solid sliding past each other.

剪切模量：固体抵抗层间相对滑动的能力的量度。

The **bulk modulus** measures the resistance of a solid (or a liquid) to a volume change.

体积模量：固体（或液体）抵抗体积变化的能力的量度。

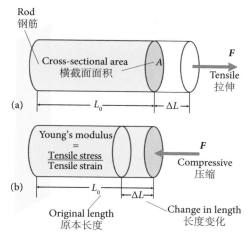

ELASTICITY IN LENGTH

A long bar clamped at one end is stretched (a) or compressed (b) by an amount ΔL under the action of a force F.

$$\text{Young's modulus} = \frac{\text{Tensile (compressive) stress}}{\text{Tensile (compressive) strain}}$$

杨氏模量=拉（压）应力/拉（压）应变

长度弹性

一个长条一端被固定，当受到力F的作用后被拉伸(a)或压缩(b)了ΔL。

ELASTICITY OF SHAPE

A shear deformation in which a rectangular block is distorted by two forces of equal magnitude but opposite directions applied to two parallel faces.

Shear modulus= Shear stress/Shear strain
剪切模量=切应力/切应变

形状弹性

剪切形变是一个长方体的相互平行的两个平面受到大小相等、方向相反的力作用而产生的形变。

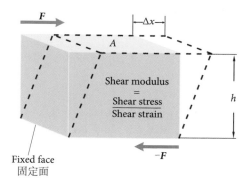

VOLUME ELASTICITY

This cube is compressed on all six sides by forces normal to its faces.

$$\text{Bulk modulus} = \frac{\text{Volume stress}}{\text{Volume strain}}$$

体积模量=体积应力/体积应变

体积弹性

该立方体的六个面都受到垂直于所在平面的力的作用。

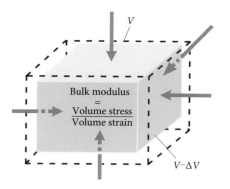

PRESSURE
压强

TERMS	术语
liquid	液体
fluid	流体
pressure	压强
atmospheric pressure	大气压
absolute pressure	绝对压强
gauge pressure (overpressure)	表压（过压）
normal force	法向力
upward force	向上的力
downward force	向下的力
force per unit area	单位面积受力
piston	活塞
variation of pressure with depth	压力随深度变化
open-tube manometer	开口压力计
mercury barometer	水银气压计
column of mercury	水银柱
altitude	海拔
density	密度
mass per unit volume	单位体积质量
open to atmosphere	通大气

LANGUAGE OF PHYSICS 物理学用语

The **pressure in a fluid** is the force per unit area that the fluid exerts on any surface.

流体的压强：流体对任意表面单位面积上施加的作用力。

Atmospheric pressure is the pressure within the atmosphere of the Earth caused by the weight of air above the measurement point.

大气压：地球大气层内的压强，是由测量点以上空气的重量引起的。

Gauge pressure is the pressure indicated on a pressure measuring gauge. It is equal to the absolute pressure minus normal atmospheric pressure.

表压：压力测量表测得的压力，数值等于绝对压力减去标准大气压。

Absolute pressure is zero-referenced against a perfect vacuum, using an absolute scale, so it is equal to gauge pressure plus atmospheric pressure.

绝对压强：以绝对真空为起点并利用绝对标度测量得到的压强，数值等于表压加大气压。

Barometer is an instrument that measures atmospheric pressure.

气压计：一种测量大气压的仪器。

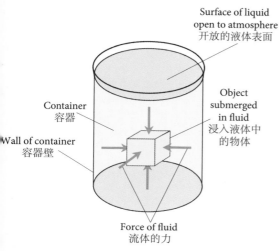

PRESSURE

The force of a fluid on a submerged object at any point is perpendicular to the surface of the object. The force of the fluid on the walls of the container is perpendicular to the walls at all points.

压强

流体对浸入其中的物体上任意一点施加的力均垂直于该物体的表面。流体对容器壁上任意一点施加的力均垂直于容器壁。

MANOMETER

The open-tube manometer is a simple device for measuring pressure. The pressure p is called the absolute pressure The difference $p - p_0$ is called the gauge pressure

压力计

开口压力计是一种测量压强的简单仪表。压强 p 被称为绝对压强，p 与 p_0 之间的差值叫作表压。

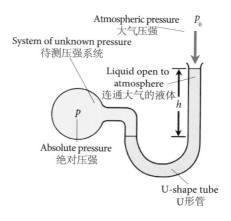

BAROMETER

The barometer, invented by Evangelista Torricelli, is another instrument used to measure pressure.

气压计

气压计是埃万杰利斯塔·托里拆利（Evangelista Torricelli）发明的，是另一种测量压强的仪器。

FLUID STATICS
流体静力学

TERMS	术语
fluid	流体
buoyant force	浮力
floating object	浮体
submerged object	浸没的物体
totally submerged object	完全浸没的物体
partially submerged object	部分浸没的物体
displaced volume of fluid	排水体积
Pascal's law	帕斯卡定律
hydraulic press	液压
hydraulic ramp	液压车桥
Archimedes' principle	阿基米德定律

LANGUAGE OF PHYSICS
物理学用语

Fluid statics or **hydrostatics** is the study of fluids at rest.

流体静力学：一门研究静止流体的学科。

A **fluid** is a collection of molecules that are randomly arranged and held together by weak cohesive forces and forces exerted by the walls of a container. Liquids and gases are fluids.

流体：由随机排列的分子构成，被非常微弱的黏滞力及容器壁施加的力聚合在一起。液体和气体都是流体。

Buoyant force is the upward force that a fluid exerts on an object submerged in it. The magnitude of the buoyant force always equals the weight of the fluid displaced by the object.

浮力：流体对浸在其中的物体施加的方向朝上的力。浮力的大小等于物体排开的流体重力的大小。

Pascal's law
When pressure is applied to an enclosed fluid, the pressure is transmitted undiminished to every point in the fluid and to every point on the walls of the container.

帕斯卡定律
对封闭的流体施加压强，压强会无损失地传递到流体和容器壁的每个点上。

Archimedes' principle
Any body completely or partially submerged in a fluid is buoyed up by a force equal to the weight of the fluid displaced by the body.

阿基米德定律
任一物体完全或部分浸入流体中时会受到一个向上的浮力，该力的大小等于物体排开的流体重力的大小。

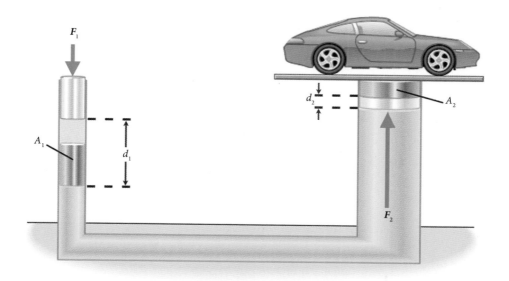

HYDRAULIC RAMP

Diagram of a hydraulic ramp. Since the increase in pressure is the same on the left and right sides, a small force F_1 at the left produces a much larger force F_2 at the right.

液压车桥

液压车桥图示。由于左右两边的压强会等量增加，因此在左边施加一个较小的力F_1时，右边会产生一个较大的力F_2。

If the weight of an object is greater than the buoyant force, the object will sink.

If the weight of an object is less than the buoyant force, the object will float.

If the weight of an object and the buoyant force are equal, the object will be suspended in the water.

当物体的重力大于浮力时，物体将下沉。

当物体的重力小于浮力时，物体将上浮。

当物体的重力等于浮力时，物体将悬浮在水中。

FLUID DYNAMICS
流体动力学

TERMS	术语
fluid	流体
fluid in motion	流动中的流体
nonviscous fluid	非黏滞流体
incompressible fluid	不可压缩流体
ideal fluid	理想流体
real fluid	真实流体
flow	流
steady flow	稳恒流
laminar flow	层流
smooth path	光滑路径
streamline	流线
set of streamlines	流线簇
flowtube	流管
pipe	管道
equation of continuity	连续性方程
Bernoulli's equation	伯努利方程式
volume flux (flow rate)	体积通量（流量）

LANGUAGE OF PHYSICS   物理学用语

Equation of continuity
The flow rate through the pipe (or the product of the cross-sectional area and the fluid speed at all points along the pipe) is constant for an incompressible fluid.

连续性方程
对于不可压缩流体而言，通过管道的流速（或管道的横截面面积与流体中各点沿管道流速的乘积）是一个常量。

Bernoulli's equation
The sum of the pressure, kinetic energy per unit volume, and gravitational potential energy per unit volume has the same value at all points along a streamline.

伯努利方程式
压强、单位体积的动能、单位体积的重力势能三者之和在流线中各点都相等。

A **streamline** is the path taken by a fluid particle under steady flow.

流线：流体中的粒子在稳恒流下经过的路径。

A **flowtube** is a set of streamlines.

流管：一组流线。

A fluid is **incompressible** if its density remains constant as pressure changes.

不论压强如何变化，流体的密度始终是个常量，那么该流体就是**不可压缩**的。

In **steady flow** the velocity of the fluid at each point remains constant in time.

稳恒流中，流体中任一点的速度不随时间变化而变化。

力学 91

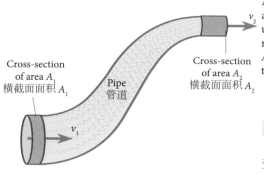

Path taken by a fluid particle
流体中的粒子经过的路径

Fluid particle
流体粒子

Velocity of the fluid particle
流体粒子的速度

Streamline
流线

Set of streamlines = Flowtube
流线簇 = 流管

FLOWTUBE

The blue lines represent a set of streamlines. A particle at B follows one of these streamlines, and its velocity is tangent to the streamline at each point along its path.

流管

图中蓝色的线条代表流线簇。一个在 B 点的粒子沿着其中某一条流线运动，它在其路径上任意一点的速度与流线相切。

EQUATION OF CONTINUITY

A fluid moving with streamlined flow through a pipe of varying cross-sectional area. The volume of fluid flowing through A_1 in a time interval Δt must equal the volume flowing through A_2 in the same time interval. Therefore, according to the equation of continuity, $A_1 v_1 = A_2 v_2$.

Cross-section of area A_1
横截面面积 A_1

Pipe
管道

Cross-section of area A_2
横截面面积 A_2

连续性方程

流体随着不定常流穿过一条横截面不断变化的管道。在相等的时间间隔 Δt 内，流过 A_1 的流体体积必然等于流过 A_2 的流体体积。因此，根据连续性方程得出，$A_1 v_1 = A_2 v_2$。

The National Nature Park "Podilski Tovtry" is located in the southern region of western Ukraine. The smaller the cross section of the river Dniester, the higher the speed of its water.

乌克兰的波多斯基国家公园位于乌克兰西南部地区。德涅斯特河横截面面积越小，其水流速度越大。

VISCOSITY
黏性

TERMS	术语
viscous fluid	黏性流体
layer of fluid	流层
internal friction or viscosity	内摩擦或黏滞现象
friction between layers	层间摩擦
flow	流动
nonsteady or unsteady flow	非稳恒流
turbulent flow	湍流
irregular flow	不规则流
irrotational flow	无旋流
smooth path	光滑路径
whirlpool region	旋涡区
coefficient of viscosity	黏度系数
speed gradient	速度梯度
Poiseuille's equation	泊肃叶方程

LANGUAGE OF PHYSICS   物理学用语

The resistance of a fluid to shearing motion is a form of internal friction called **viscosity**.

黏性：流体的一种内摩擦力，以抵抗流体剪切运动。

The **viscous force** or **internal friction** is the resistance of two adjacent layers of the fluid moving relative to each other.

黏滞力或内摩擦力：两个相邻流层间为抵抗相对运动而产生的力。

The flow is **laminar** (or **steady**) if each particle of the fluid follows a smooth path, so that the paths of different particles never cross each other.

如果流体中的粒子沿着平滑路径流动，那么其流动是分层的（或稳定的），此时不同粒子的路径永不相交。

Turbulent flow is an irregular flow characterized by small whirlpool regions.

湍流：一种不规则的流动，湍流中会出现小的旋涡区。

LAMINAR AND TURBULENT FLOW

(a) Laminar flow occurs in layers without mixing. Viscosity causes drag between layers as well as with the fixed surface.

(b) An obstruction in the vessel produces turbulence. Turbulent flow mixes the fluid. There is more interaction, greater heating, and more resistance than in laminar flow.

层流和湍流

（a）层流分层出现，且不同层之间不会出现混合。黏滞力阻碍了层间的运动，也起到了固定层表面的作用。

（b）容器中的突起会引发湍流，使流体混合。与层流相比，湍流中的相互作用更多，产生的热量更高，且阻力更大。

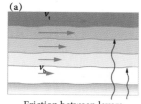
(a)

(b)

Friction between layers
层间摩擦

SPEED OF A FLUID

(a) If fluid flows in a tube that has negligible resistance, the speed is the same all across the tube.
(b) When a viscous fluid flows through a tube, its speed at the walls is zero, increasing steadily to its maximum at the center of the tube.

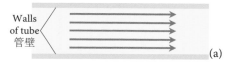
Walls of tube
管壁
(a)

Maximum speed
最大速率
(b)

流体速度

（a）如果管壁的阻力可忽略，则经过管道各处的流速相等。

（b）当黏性流体在管内流动时，管壁处速度为0，距离管壁越远，流速越快，管中心位置速度最大。

A car undergoes an aerodynamic test in a wind tunnel. The streamlines in the air flow are made visible by smoke particles.

一辆汽车正在风洞中进行气动力试验。气流流线由于烟雾粒子的存在而变得显而易见。

SIMPLE HARMONIC MOTION
简谐运动

TERMS 术语

periodic (oscillatory) motion	周期（振荡）运动
oscillation	振动
number of oscillation	振动次数
displacement from equilibrium	偏离平衡位置的位移
amplitude	振幅
phase constant (or phase angle)	相位常量（或相位角）
phase	相位
period	周期
frequency	频率
angular frequency	角频率
simple harmonic oscillator	简谐振子
mass attached to a spring	附在弹簧上的物体

LANGUAGE OF PHYSICS 物理学用语

Motion that repeats itself in equal intervals of time is called **periodic motion**.

周期运动：在相等时间间隔后重复的运动。

Simple harmonic motion is the repetitive back-and-forth motion about an equilibrium position when the force on a body is proportional to its displacement from equilibrium.

简谐运动：当物体所受的力与其偏离平衡位置的位移成正比时，物体在平衡位置附近进行的往复运动。

A **cycle** is one complete oscillation.

（循环）周期：一个完整的振动过程。

The **displacement** is the distance a vibrating body moves from its equilibrium position.

位移：振动的物体偏离平衡位置的距离。

The **amplitude** is the maximum displacement of the particle from the equilibrium position.

振幅：质点偏离平衡位置的最大位移。

The **period** is the time it takes the particle to go through one full cycle.

（时间）周期：质点经历一个完整循环所用的时间。

The **frequency** is the number of oscillations in unit time. The **frequency** is the reciprocal of the period.

频率：单位时间内的振动次数。频率是周期的倒数。

The **angular frequency** is the angular velocity of the vibrating system.

角频率：振动系统的角速度。

CHARACTERISTICS OF SIMPLE HARMONIC MOTION

Displacement versus time for a particle undergoing simple harmonic motion. The amplitude of the motion is A, and the period is T.

简谐运动的特征

做简谐运动的物体位移随时间的变化。运动的振幅为 A，周期为 T。

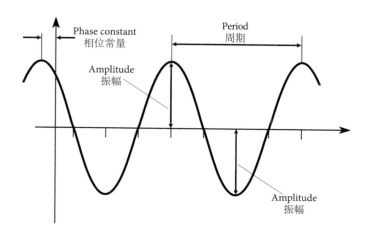

DISPLACEMENT, VELOCITY AND ACCELERATION

An oscillating mass attached to a spring moves in simple harmonic motion.

(a) When the mass is displaced up from equilibrium position, the displacement is positive and the acceleration is negative.

(b) At the equilibrium position, displacement equals zero, the acceleration is zero but the speed is a maximum.

(c) When the displacement is negative, the acceleration is positive.

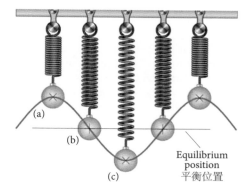

位移、速度和加速度

一个附在弹簧上的物体做简谐运动。

（a）当物体处在平衡位置以上时，位移为正，加速度为负。

（b）在平衡位置处，位移等于 0，加速度等于 0，但速率达到最大值。

（c）当位移为负时，加速度为正。

PENDULUM
摆

TERMS	术语
mechanical system	机械系统
pendulum	摆
simple pendulum	单摆
physical (compound) pendulum	物理（复）摆
torsional pendulum	扭摆
spring pendulum	弹簧摆
reversible pendulum	可逆摆
elastic pendulum	弹性摆
ballistic pendulum	弹道摆
point mass	质点
light string	轻质悬线
length of string	悬线长度
restoring force	回复力
gravitational force	重力
center of mass	质心
pivot point	支点

LANGUAGE OF PHYSICS   物理学用语

A **simple pendulum** is a point mass suspended by a light string that moves in an oscillatory motion under the action of gravity.

单摆：悬挂在轻质悬线下的一个质点，在重力的作用下会产生振动。

The **period and frequency of a simple pendulum** depend only on the length of the string and the value of the gravitational field strength.

单摆的周期和频率：仅由悬线长度和重力场的强度值决定。

A **physical pendulum** is a hanging object which oscillates about a fixed axis that does not pass through its center of mass.

物理摆：绕不过质心的固定轴振动的一个悬垂物体。

A **torsional pendulum** consists of a disk-like mass suspended from a thin rod or wire. When the mass is twisted about the axis of the wire, the wire exerts a restoring torque on the mass, tending to rotate it back to its original position.

扭摆：一个悬挂在细杆或细线上的盘状物。当物块绕细线所在的轴发生扭转时，细线会对物体产生一个回复力矩，使其转回初始位置。

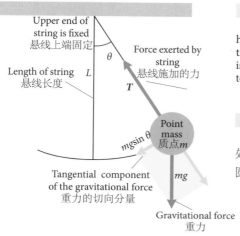

SIMPLE PENDULUM

The simple pendulum oscillates in simple harmonic motion about the equilibrium position ($\theta = 0$) when θ is small. The restoring force is $mg\sin\theta$, the component of the weight tangent to the circle.

单摆

当 θ 很小时，单摆围绕平衡位置（$\theta=0$ 处）做简谐运动。回复力等于 $mg\sin\theta$，即与圆相切的重力的分量。

PHYSICAL PENDULUM

The physical pendulum consists of a rigid body pivoted at the point O, which is not at the center of mass. At equilibrium, the weight vector passes through O, corresponding to $\theta = 0$. The restoring torque about O when the system is displaced through an angle θ is $mgd\sin\theta$.

复摆

复摆由一个固定在 O 点的刚体完成，且 O 点不与质心重合。平衡时，重力向量经过 O 点，此时 $\theta=0$。当该装置偏移角度 θ 时，O 点的回复力矩大小等于 $mgd\sin\theta$。

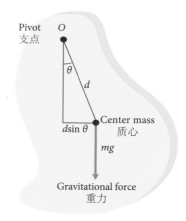

The motion of a simple pendulum captured with multiple exposure photography.

多次曝光拍摄的单摆运动。

DAMPED OSCILLATION
阻尼振动

TERMS	术语
ideal system	理想系统
real system	真实系统
dissipative force	耗散力
retarding force	减速力
restoring force	回复力
damped harmonic motion	阻尼谐振动
oscillator	振子
damped oscillator	阻尼振子
undamped oscillator	无阻尼振子
envelope of oscillatory curve	振动曲线包络
exponential factor	指数因子
natural frequency	固有频率
oscillation	振动
underdamped oscillation	欠阻尼振动
critically damped oscillation	临界阻尼振动
overdamped oscillation	过阻尼振动
damping rate	阻尼率
logarithmic decrement	对数衰减量/率
coefficient of damping (damping ratio)ζ	阻尼系数（阻尼比）ζ
coefficient of resistance of medium	介质阻尼系数
quality factor	品质因子

LANGUAGE OF PHYSICS
物理学用语

A **damped oscillator** is a system where there is a retarding force (proportional to velocity) as well as a restoring force (proportional to displacement).

阻尼振子：同时存在减速力（与速度成正比）和回复力（与位移成正比）的系统。

A damped harmonic oscillator can be:

阻尼谐振子可为以下三种：

Overdamped ($\zeta > 1$): The system returns (exponentially decays) to steady state without oscillating.

过阻尼($\zeta > 1$)：系统没有振动地回到（指数衰减到）稳态。

Critically damped ($\zeta = 1$): The system returns to steady state as quickly as possible without oscillating.

临界阻尼($\zeta = 1$)：系统不经历振动而尽快回到稳态。

Underdamped ($\zeta < 1$): The system oscillates (with a slightly different frequency than the undamped case) with the amplitude gradually decreasing to zero.

欠阻尼($\zeta < 1$)：系统做振幅逐渐减小的振动（频率与无阻尼振动略微不同），直至不动。

力学

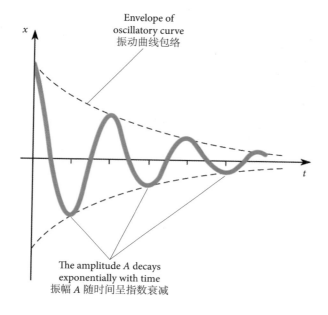

DAMPED OSCILLATION

Graph of displacement versus time for a damped oscillator. The retarding force is proportional to the speed and acts in the direction opposite to the motion.

阻尼振动

阻尼振子的位移-时间图。减速力与振子的速率成正比，与其运动方向相反。

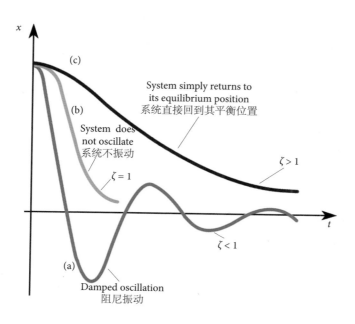

TYPES OF DAMPED OSCILLATIONS

Plots of displacement versus time for
(a) an underdamped oscillator,
(b) a critically damped oscillator,
(c) an overdamped oscillator.

阻尼振动的类型

位移-时间曲线：
(a) 欠阻尼振子；
(b) 临界阻尼振子；
(c) 过阻尼振子。

FORCED OSCILLATION
受迫振动

TERMS 术语

energy loss	能量损耗
damped system	阻尼系统
external force	外力
driving force	驱动力
periodic force	周期力
positive work	正功
forced oscillator	受迫振子
steady-state conditions	稳态条件
resonance	共振
resonance frequency	共振频率
natural frequency	固有频率
resonance curve	共振曲线
resonant vibration	共振
damping coefficient	阻尼系数

LANGUAGE OF PHYSICS
物理学用语

A **forced oscillator** is a damped oscillator driven by an external force that varies periodically.

受迫振子：受周期性外力驱动的阻尼振子。

Resonance is the dramatic increase in amplitude near the natural frequency. At resonance the applied force is in phase with the velocity and the power transferred to the oscillator is a maximum.

共振：接近固有频率时，振幅急剧增加的现象。引发共振所施加的力与速度同相位，且转移给振子的能量最大。

RESONANCE CURVE

Graph of amplitude versus frequency for a damped oscillator when a periodic driving force is present. When the frequency of the driving force equals the natural frequency, resonance occurs. The shape of the resonance curve depends on the size of the damping coefficient.

共振曲线

周期性驱动力作用下阻尼振子的振幅-频率图。当驱动力的频率等于固有频率时，发生共振。共振曲线的形状取决于阻尼系数的大小。

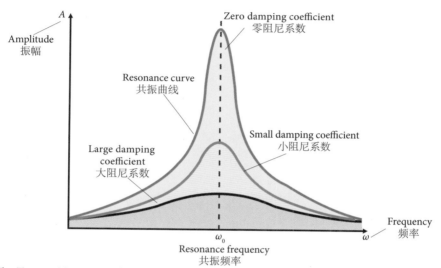

The Tacoma Narrows Bridge in Washington was destroyed by resonant vibration (1940). The bridge collapsed because turbulences generated by the wind blowing through the bridge structure occurred at a frequency that matched the natural frequency of the structure.

美国华盛顿州的塔科马海峡吊桥在1940年毁于共振。风吹过桥体结构时产生了湍流，且该湍流的频率恰好与桥体结构的固有频率相同，最终导致了桥梁的坍塌。

WAVE MOTION
波动

TERMS 术语

wave	波
travelling wave	行波
transverse wave	横波
longitudinal wave	纵波
mechanical wave	机械波
sinusoidal wave	正弦波
plane wave	平面波
spherical wave	球面波
wave front	波前
disturbance	扰动
source of disturbance	干扰源
disturbed medium	被干扰介质
direction of propagation	传播方向
wavelength	波长
frequency	频率
amplitude	振幅
displacement	位移
phase constant	相位常量
period	周期
wave equation	波动方程
wave function	波函数
wave speed	波速
angular frequency	角频率
angular wave number	角波数
crest wave	波峰
trough	波谷

LANGUAGE OF PHYSICS 物理学用语

A **wave** is a propagation of a disturbance through a medium.

波：介质中扰动的传播。

The **wavelength** is the distance in the direction of propagation in which the wave repeats itself.

波长：在波传播方向上，波形重复出现的距离。

The **frequency of a periodic wave** is the time rate at which the disturbance repeats itself.

周期波的频率：扰动重复出现的时间速率。

The **wave function** represents the value of the function at any point and at any moment of time.

波函数：表示函数在任意一点任意时刻对应的值。

力学 103

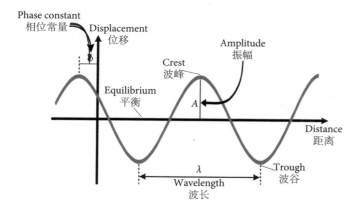

TRAVELLING WAVE

The travelling sinusoidal wave moves to the right. The wavelength λ of a wave is the distance between adjacent crests or adjacent troughs.

行波

行进中的正弦波向右移动。波的波长λ等于相邻两个波峰或波谷之间的距离。

FREQUENCY AND PERIOD

The frequency of a sinusoidal wave is the inverse of the period.

频率和周期

正弦波的频率是周期的倒数。

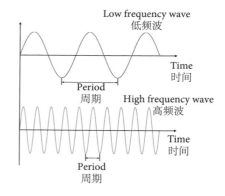

LONGITUDINAL AND TRANSVERSE WAVES

A longitudinal wave is one in which the particles of the medium move in a direction parallel to the direction of the wave velocity.

A transverse wave is one in which the particles of the medium move in a direction perpendicular to the direction of the wave velocity.

纵波和横波

纵波：介质中的粒子沿平行于波速方向运动的波。

横波：介质中的粒子沿垂直于波速方向运动的波。

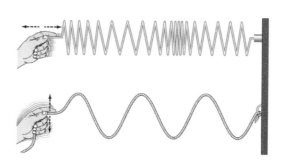

INTERFERENCE of WAVE
波的干涉

TERMS 术语

travelling wave	行波
wave combination	波的混合
superposition principle	叠加原理
resultant wave function	合成波函数
linear wave	线性波
nonlinear wave	非线性波
interference pattern	干涉图样
interference	干涉
constructive interference	相长干涉
destructive interference	相消干涉
phase angle	相位角
waveform	波形

LANGUAGE OF PHYSICS 物理学用语

The **superposition principle** says that when two or more waves move through a medium, the resultant wave function equals the algebraic sum of the individual wave functions.

叠加原理：两个或两个以上的波在介质中传播时，合成波函数等于各个波函数的代数和。

Constructive interference
When two interfering waves are in phase with each other (phase angle = 0) the amplitude of the combined wave is a maximum.

相长干涉
同相位（即相位角=0）的两列波相干时，合成波的振幅最大。

Destructive interference
When two interfering waves are 180° out of phase with each other and of equal amplitude then the amplitude of the combined wave is zero.

相消干涉
振幅相等、相位相差180°的两列波相干时，合成波的振幅为0。

The **phase angle** is a measure of how far one wave is displaced in the direction of propagation from another wave.

相位角：两列波在传播方向上相距远近的量度。

The **amplitude** is the maximum value of the displacement.

振幅：位移的最大值。

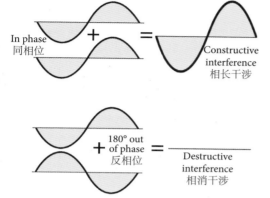

CONSTRUCTIVE AND DESTRUCTIVE INTERFERENCE

The interference may be constructive, when the individual displacements are in the same direction, or destructive when the displacements are in opposite directions.

In phase 同相位
Constructive interference 相长干涉

180° out of phase 反相位
Destructive interference 相消干涉

相长干涉与相消干涉

两列波的位移同向时，干涉相长；位移反向时，干涉相消。

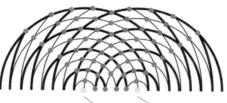

INTERFERENCE PATTERN

When two waves combine in space, they interfere to produce a resultant wave pattern.

Source 1 波源 1
Source 2 波源 2

干涉图样

当两列波在空间上混合在一起时，它们因干涉而产生一个合成波谱。

Interference pattern produced by outward spreading waves from two drops of water falling into a pond. Two travelling waves can pass through each other without being destroyed or even altered.

两滴水掉落在水塘后形成的两列波向外传播时产生的干涉图样。两列行波能够穿过彼此，而不会遭到破坏或改变。

SOUND WAVE
声波

TERMS	术语
wave	波
audible (sound) wave	声波
infrasonic wave	次声波
ultrasonic wave	超声波
longitudinal wave	纵波
compressible medium	可压缩介质
speed of sound wave	声波速度
intensity of wave	波的强度
compressed (condensation) region	压缩区域（密部）
low-pressure (rarefaction) region	低压区域（疏部）
displacement amplitude	位移振幅
pressure amplitude	压强幅度
variation in pressure	压强变化
Doppler effect	多普勒效应
source	波源
observer	观测者

LANGUAGE OF PHYSICS   物理学用语

A **sound wave** is longitudinal and travel through a compressible medium with a speed that depends on the compressibility and inertia of that medium.

声波：在可压缩介质中传播的一种纵波，其波速取决于介质的可压缩性和惯性。

The **Doppler effect** means the change in the wavelength and hence the frequency of a sound caused by the relative motion between the source and the observer.

多普勒效应：波源和观测者之间存在相对运动时，声波的波长和频率会发生变化。

When a **moving source approaches a stationary observer** the observed frequency is higher than the emitted frequency of the source.

当**波源向着静止的观测者运动**时，观测到的波的频率会高于波源的发射频率。

When a **moving source recedes from a stationary observer**, the observed frequency is lower than the emitted frequency of the source.

当**波源远离静止的观测者**时，观测到的波的频率会低于波源的发射频率。

The **intensity of a wave** is the energy of a wave that passes a unit area in unit time.

波的强度：一列波在单位时间通过单位面积的能量。

PROPAGATION OF SOUND WAVE / 声波的传播

A sound wave is a longitudinal wave that can be propagated in a solid, liquid, or gas (transverse waves can also propagate in solids).

声波是一种能够在固体、液体或气体中传播的纵波（横波也能在固体中传播）。

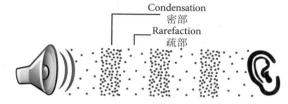

Condensation 密部
Rarefaction 疏部

SOUND LEVELS FOR SOME SOURCES / 一些声源的声级

Source of sound	Sound level (dB) 声级（dB）	声源
jet engine	150	喷气发动机
pain threshold	130	痛阈
rock concert	120	摇滚音乐会
pneumatic drill	100	风钻
street traffic	80	路面交通
conversation	60	对话
birdsong	40	鸟鸣
whisper	20	耳语
falling leaves	10	落叶
hearing threshold	0	听阈

THE DOPPLER EFFECT / 多普勒效应

A source moving with a speed v toward a stationary observer A and away from a stationary observer B. Observer A hears an increased frequency, and observer B hears a decreased frequency.

声源以速度 v 靠近静止的观测者 A，远离静止的观测者 B。观测者 A 听到的声音频率增大，而观测者 B 听到的声音频率减小。

The Doppler effect is used in police radar systems to measure the speed of motor vehicles.

警察的雷达系统利用多普勒效应来测量机动车的速度。

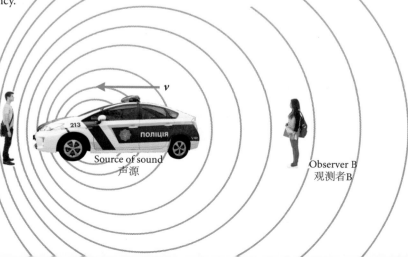

Observer A 观测者A
Source of sound 声源
Observer B 观测者B

SHOCK WAVE
冲击波

TERMS	术语
shock wave	冲击波
sonic boom	音爆
supersonic speed	超声速
subsonic speed	亚声速
loud explosion	响亮的爆炸声
source speed	波源速度
wave speed	波速
wave front	波前
spherical wave front	球面波前
conical wave front	锥面波前
envelope of wave	波的包络面
cone	锥
surface of cone	锥表面
half-apex angle	半顶角
pressure variation	压强变化
Mach number	马赫数

LANGUAGE OF PHYSICS 物理学用语

The conical wave front produced when the speed of a sound source exceeds the wave speed (supersonic speed) is known as a **shock wave.**

冲击波：当声源的速度超过波速（超声速）时产生的一种锥形波前。

SHOCK WAVE FORMATION

(a) A source moves slower than sound.
(b) The source speed equals the wave speed.
(c) A representation of a shock wave produced when a source moves with a speed greater than the wave speed in that medium.

冲击波的形成

（a）波源运动速度小于声速。
（b）波源运动速度等于波的传播速度。
（c）波源运动速度大于介质中波的传播速度，产生冲击波。

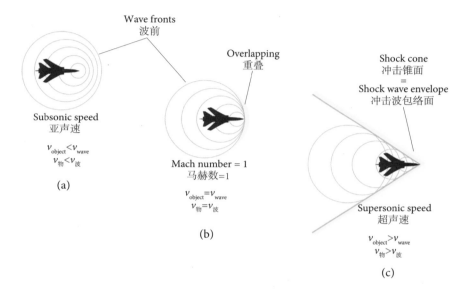

Jet airplanes travelling at supersonic speeds produce shock waves, which are responsible for the loud explosion, or "sonic boom" that may be heard. The shock wave carries a great deal of energy concentrated on the surface of the cone, with correspondingly large pressure variations.

喷气式飞机以超声速飞行时会产生冲击波，并伴随响亮的爆炸声，即可能被听到的"音爆"。冲击波携带着大量的能量，该能量集中在波前锥面上，并导致巨大的压强变化。

STATIONARY WAVE
驻波

TERMS	术语
mode of vibration	振动模式
path length	路径长度
node	波节
antinode	波腹
wave function for stationary wave	驻波的波函数
position of node	波节位置
position of antinode	波腹位置
one quarter of a period	1/4 周期
half of a period	半周期
normal mode	简正模式
fundamental frequency	基频
first (second...) harmonic	一次（二次/……）谐波
wave speed	波速
complex wave	复合波
beat	拍波
damped beat	阻尼拍波
zero beat	零拍波

LANGUAGE OF PHYSICS 物理学用语

Stationary waves are formed from the superposition of two harmonic waves having the same frequency, amplitude, and wavelength but travelling in opposite directions.

驻波：由两列频率相等、振幅相等、波长相等，但是传播方向相反的谐波叠加而成。

An **antinode** is a point of maximum amplitude.

波腹：振幅最大的点。

A **node** is a point of zero amplitude.

波节：振幅为零的点。

STATIONARY WAVE

A stationary wave is a stationary vibration pattern formed by the superposition of two waves of the same frequency travelling in opposite directions.

驻波

驻波是由频率相同、传播方向相反的两列波叠加而成的静止的振动模式。

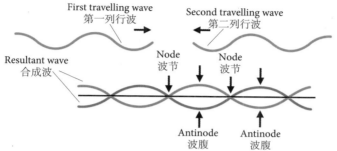

NODES AND ANTINODES

Adjacent antinodes are separated by $\lambda/2$. The nodes are also spaced by $\lambda/2$. The distance between a node and an adjacent antinode is $\lambda/4$.

波节和波腹

相邻波腹之间的距离为 $\lambda/2$，相邻波节之间的距离也是 $\lambda/2$，波节和相邻波腹之间的距离是 $\lambda/4$。

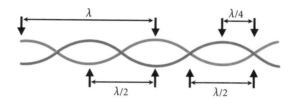

These waves developed on a small creek just as it entered Lake Michigan on the beach at Warren Dunes State Park. The water was flowing clearly and smoothly until it reached the lake, when these bumps would appear. They didn't flow forward like a regular wave; they just stayed in the same place.

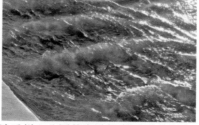

一条小溪刚刚汇入沃伦沙丘州立公园的密歇根湖时形成的波。水流原本清澈、平滑，在汇入密歇根湖时才出现起伏。这些起伏不像普通的波一样向前传播，而是停留在原地。

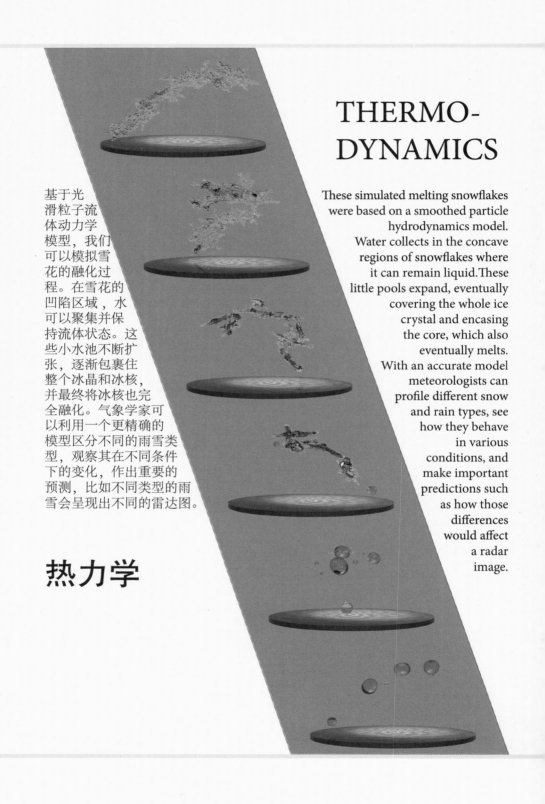

THERMO-DYNAMICS

These simulated melting snowflakes were based on a smoothed particle hydrodynamics model. Water collects in the concave regions of snowflakes where it can remain liquid. These little pools expand, eventually covering the whole ice crystal and encasing the core, which also eventually melts. With an accurate model meteorologists can profile different snow and rain types, see how they behave in various conditions, and make important predictions such as how those differences would affect a radar image.

基于光滑粒子流体动力学模型，我们可以模拟雪花的融化过程。在雪花的凹陷区域，水可以聚集并保持流体状态。这些小水池不断扩张，逐渐包裹住整个冰晶和冰核，并最终将冰核也完全融化。气象学家可以利用一个更精确的模型区分不同的雨雪类型，观察其在不同条件下的变化，作出重要的预测，比如不同类型的雨雪会呈现出不同的雷达图。

热力学

THERMODYNAMICS

Thermodynamics is a branch of physics concerned with heat and temperature and their relation to energy and work. It defines macroscopic variables, such as internal energy, entropy, and pressure, that provide partial descriptions of a body of matter or radiation.

PARTS OF THERMODYNAMICS

Classical thermodynamics is the description of the states of thermodynamic systems at or near equilibrium, that uses macroscopic, measurable properties. It is used to model exchanges of energy, work and heat based on the laws of thermodynamics.

Statistical thermodynamics supplements classical thermodynamics with an interpretation of the microscopic interactions between individual particles or quantum-mechanical states.

Chemical thermodynamics is the study of the interrelation of energy with chemical reactions or with a physical change of state within the confines of the laws of thermodynamics.

Equilibrium thermodynamics is the systematic study of transformations of matter and energy in systems when they are close to approach equilibrium.

热力学

热力学：研究热和温度及它们与能量和功的关系的物理学分支。热力学定义了宏观变量，如内能、熵和压力，这些变量可对物体或辐射体进行部分描述。

热力学分类

经典热力学：用宏观可测量的性质来描述处于或接近平衡的热力学系统的状态。根据热力学定律，经典热力学可以用来模拟能量、功和热的交换。

统计热力学：经典热力学的补充，用于解释个别粒子之间或量子力学状态之间的微观相互作用。

化学热力学：在热力学定律的范围内，研究能量与化学反应或物态变化之间的相互关系。

平衡热力学：系统地研究系统在接近平衡状态时物质和能量的转换。

TEMPERATURE
温度

TERMS 术语

thermal contact	热接触
thermal equilibrium	热平衡
heat	热
zeroth law of thermodynamics	热力学第零定律
thermometer	温度计
constant-volume gas thermometer	定容气体温度计
mercury-in-glass thermometer	水银温度计
alcohol thermometer	酒精温度计
infrared thermometer	红外线温度计
temperature scale	温标
Kelvin temperature scale	开氏温标
Celsius temperature scale	摄氏温标
Fahrenheit temperature scale	华氏温标
thermodynamic temperature scale	热力学温标
water / water vapor / ice	水/水蒸气/冰
triple point	三相点
boiling point	沸点
freezing (melting) point	冰点
ice point of water	水的冰点
steam point of water	水的汽点
absolute zero	绝对零度

LANGUAGE OF PHYSICS 物理学用语

Temperature is a measure of the average kinetic energy of the particles that make up a body.

温度：组成物体的粒子的平均动能的量度。

The **Celsius temperature scale** uses 0 ℃ for the melting point of ice and 100 ℃ for the boiling point of water at normal atmospheric pressure.

摄氏温标：标准大气压下，冰的熔点为 0 ℃，水的沸点为100 ℃。

The **Kelvin temperature scale** is based on the pressure of an ideal gas. Absolute zero (0 K) is the nominal temperature at which gas pressure becomes zero.

开氏温标：以理想气体的压强为基础，定义绝对零度（0 K）为气体压强为零时的标称温度。

A **thermometer** is a device for measuring the temperature of a body.

温度计：测量物体温度的一种仪器。

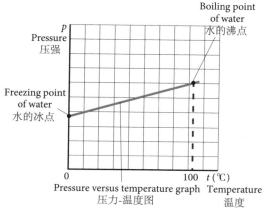

PRESSURE VERSUS TEMPERATURE

A graph of pressure versus temperature taken with a constant-volume gas thermometer. The dots represent known reference temperatures (the freezing point and the boiling point of water).

压强-温度

用定容气体温度计测得的压强-温度曲线图。圆点代表已知的参考温度（水的冰点和沸点）。

ABSOLUTE ZERO

Pressure versus temperature for dilute gases. For all gases the pressure extrapolates to zero at the unique temperature of −273.15 ℃.

绝对零度

稀薄气体的压力-温度图。将所有气体的压强外推至零将对应同一个温度值，即 −273.15 ℃。

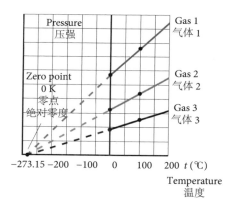

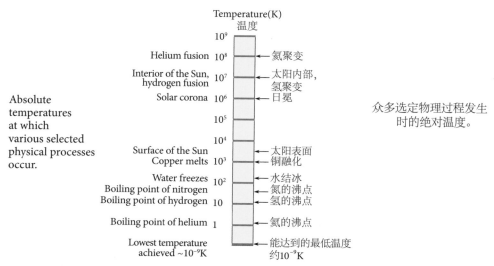

Absolute temperatures at which various selected physical processes occur.

众多选定物理过程发生时的绝对温度。

THERMAL EXPANSION of SOLID
固体热膨胀

TERMS	术语
temperature variation	温度变化
initial length	初始长度
final length	最终长度
change in length	长度变化
change in volume	体积变化
average coefficient of linear expansion	平均线膨胀系数
average coefficient of volume expansion	平均体膨胀系数
bimetallic strip	双金属条
thermostat	恒温器
steel rod	钢条
brass rod	黄铜条
thermal expansion joint	热膨胀缝

LANGUAGE OF PHYSICS
物理学用语

Thermal expansion is the tendency of matter to change in shape, area, and volume in response to a change in temperature, through heat transfer.

热膨胀：物质的形状、面积和体积随温度的变化通过热交换而发生变化的趋势。

When the temperature of an object is changed by an amount ΔT, its length changes by an amount ΔL that is proportional to ΔT and to its initial length L_0.

当物体的温度改变ΔT时，其长度改变量ΔL与ΔT及其初始长度L_0成正比。

Thermal expansion joints are used to separate sections of roadways on bridges. Without these joints, the surface would buckle due to thermal expansion on very hot days or crack due to contraction on very cold days.

热膨胀缝用于在桥梁上分隔路段。如果没有这些接缝，桥梁表面就会在酷暑时由于热膨胀而变形，或者在严寒时由于收缩而开裂。

THERMAL OSCILLATIONS IN SOLIDS

(a) At ordinary temperatures, the atoms or molecules in a solid vibrate about their equilibrium position with an amplitude of approximately 10^{-11} m. (b) As the temperature of the solid increases, the atoms vibrate with larger amplitude and the average separation between them increases.

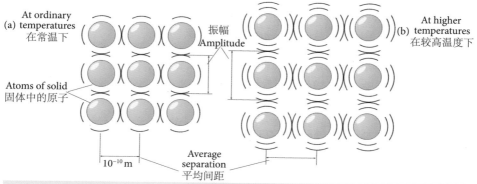

固体中的热振动

(a)在常温下，固体中的原子或分子会围绕其平衡位置振动，振幅约为10^{-11}米。(b)随着固体温度的升高，原子振动幅度增大，原子平均间距也会随之增大。

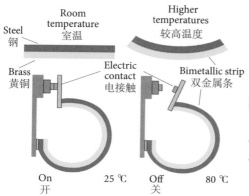

BIMETALLIC STRIP

(a) A bimetallic strip bends as the temperature changes because the two metals have different expansion coefficients. (b) A bimetallic strip used in a thermostat to break or make electrical contact.

双金属条

(a)由于两种金属的膨胀系数不同，双金属条随着温度的变化而弯曲。(b) 恒温器中用来断开或接通电路的双金属条。

An example of application of the thermal expansion joints on the roadways.

热膨胀缝在道路上的应用实例。

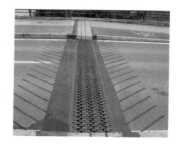

LAWS of an IDEAL GAS
理想气体定律

TERMS 术语

gas	气体
ideal gas	理想气体
low-density gas	低密度气体
gas at room temperature	室温气体
gas at atmospheric pressure	大气压下的气体
thermodynamic variable	热力学变量
temperature	温度
volume	体积
pressure	压强
equation of state	状态方程式
ideal gas law	理想气体定律
number of moles	摩尔数
Avogadro's number	阿伏伽德罗常数
molar mass	摩尔质量
universal gas constant	通用气体常数
Boltzmann's constant	玻尔兹曼常数

LANGUAGE OF PHYSICS 物理学用语

The ideal gas law

The general gas law contains Boyle's, Gay-Lussac's and Charles' law as special cases. It states that the product of the pressure and volume of a gas divided by the absolute temperature of the gas is a constant.

One **mole** of any gas is that amount of the gas that has a mass in grams equal to the atomic or molecular mass of the gas. One mole of any gas at a temperature of 0 ℃ and a pressure of one atmosphere, has a volume of 22.4 liters.

Avogadro's number N_A is the number of molecules that are contained in the amount of substance given by one mole.

Boltzmann's constant k_B is the gas constant R divided by Avogadro's number N_A.

理想气体定律

普适气体定律建立在波义耳定律、盖-吕萨克定律和查理定律这些特例的基础上。该定律表明，气体的压强和体积的乘积除以气体的绝对温度是一个常数。

1 摩尔任意气体的质量以克为单位时等于该气体的原子或分子质量。在 0 ℃且 1 个大气压环境下，1 摩尔任意气体的体积是 22.4 升。

阿伏伽德罗常数 N_A：1 摩尔物质所含的分子数。

玻尔兹曼常数 k_B 等于气体常数 R 除以阿伏伽德罗常数 N_A。

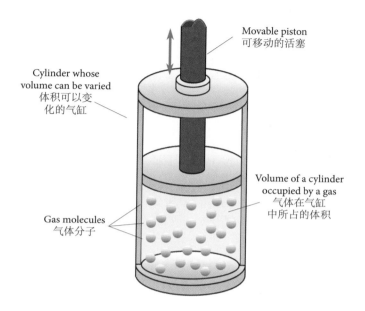

THERMODYNAMIC PARAMETERS

An ideal gas is confined to a cylinder, whose volume can be varied with a movable piston. The state of the gas is defined by any two properties, such as pressure, volume and temperature, with the third being fixed by the other two.

热力学参数

气缸内理想气体的体积随着活塞的移动而变化。气体的状态是由压力、体积和温度中的任意两个参数决定的，第三个参数也由这两个参数来确定。

THE IDEAL GAS EQUATION

Equation of state for an ideal gas.

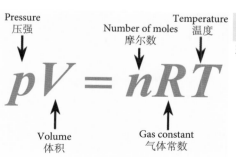

理想气体状态方程

理想气体状态方程如左侧所示。

ISOPROCESSES
等温、等压和等容过程

TERMS	术语
law	定律
Boyle's law	波义耳定律
Gay-Lussac's law	盖-吕萨克定律
Charles' law	查理定律
process	过程
isothermal process	等温过程
isobaric process	等压过程
isovolumetric (isochoric) process	等容过程
adiabatic process	绝热过程
adiabatic free expansion	绝热自由膨胀
isotherm	等温线
hyperbola	双曲线

LANGUAGE OF PHYSICS   物理学用语

An **isobaric process** is a process that takes place at a constant pressure.

等压过程：压强不变的过程。

Charles' law
The volume of a gas at constant pressure is directly proportional to the absolute temperature of the gas.

查理定律
在压强不变的情况下，气体的体积与其绝对温度成正比。

An **isochoric process** is a process that takes place at a constant volume.

等容过程：体积不变的过程。

Gay-Lussac's law
The absolute pressure of a gas at constant volume is directly proportional to the absolute temperature of the gas.

盖-吕萨克定律
在体积不变的情况下，气体的绝对压强与其绝对温度成正比。

An **isothermal process** is a process that takes place at a constant temperature.

等温过程：温度不变的过程。

Boyle's law
The product of the pressure and volume of a gas at constant temperature is equal to a constant.

波义耳定律
当温度恒定时，气体的压强和体积的乘积等于一个常数。

An **adiabatic process** is a process that occurs without an exchange of heat between the system and enviroment.

绝热过程：系统与环境之间不发生热交换的过程。

热力学 **121**

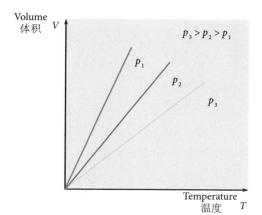

ISOTHERMAL PROCESS

A graph of pressure p versus volume V for an isothermal process that occurs at constant temperature.

等温过程

等温过程中压强 p 和体积 V 的曲线图。

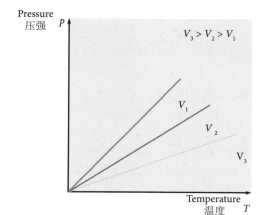

ISOBARIC PROCESS

A graph of volume V versus temperature T for an isobaric process that occurs at constant pressure.

等压过程

等压过程中体积 V 与温度 T 的曲线图。

ISOVOLUMETRIC PROCESS

A graph of pressure p versus temperature T for an isovolumetric process that occurs at constant volume.

等容过程

等容过程中压强 p 与温度 T 的曲线图。

HEAT
热

TERMS	术语
heat	热
thermal energy	热能
thermal energy transfer	热能转换
heat capacity	热容
specific heat	比热
molar specific heat	摩尔比热
calorimetry	量热法
calorimeter	量热计
latent heat	潜热
latent heat of fusion	熔化潜热
latent heat of vaporization	汽化潜热
phase change	相变
melting	熔化
boiling	沸腾
freezing	结冰
evaporation	蒸发
condensation	凝结
sublimation	升华
deposition	凝华

LANGUAGE OF PHYSICS   物理学用语

The **heat capacity** of any object is the amount of thermal energy needed to raise the temperature of the object by one degree Celsius.

物体的**热容**：物体的温度上升 1 ℃需要的热能。

The **specific heat** or **specific heat capacity**, of a substance is the heat capacity per unit mass.

物质的**比热**或**比热容**：单位质量的热容。

The **molar specific heat capacity** of a substance is the heat capacity per mole.

物质的**摩尔比热容**：每摩尔的热容。

Phases of matter include the solid phase, the liquid phase and the gaseous phase. Matter can exist in one of these phases.

物质的相包括固相、液相和气相。物质可以其中任何一种相的形式存在。

A **change of phase** is the change in a substance from one phase of matter to another. **Melting** is a change from the solid state to the liquid state. **Boiling** is a change from the liquid state to the gaseous state.

相变：物质从一种相到另一种相的转变。**熔化**是指从固态到液态的转变。**沸腾**是指从液态到气态的转变。

LATENT HEAT OF FUSION

The latent heat of fusion is the amount of heat necessary to convert 1 kg of the solid to 1 kg of the liquid.

Ice at 0 ℃
0 ℃的冰

Latent heat of fusion
熔化潜热

Water at 0 ℃
0 ℃的水

熔化潜热

熔化潜热为将1千克固体转化为1千克液体需要的热量。

LATENT HEAT OF VAPORIZATION

The latent heat of vaporization is the amount of heat necessary to convert 1 kg of the liquid to 1 kg of the gas.

Water at 100 ℃
100 ℃的水

Latent heat of vaporization
汽化潜热

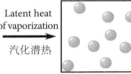

Steam at 100 ℃
100 ℃的蒸汽

汽化潜热

汽化潜热为将1千克液体转化为1千克气体需要的热量。

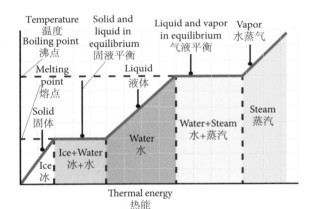

CONVERTING ICE TO STEAM

A plot of temperature versus thermal energy added when ice initially at negative temperature (in ℃) is converted to steam.

将冰转化为蒸汽

从负温度（以摄氏度为单位）的冰转化为蒸汽时的温度-热能图。

PHASE CHANGE DIAGRAM

The phase change diagram illustrates the phase changes that take place when there is a change of temperature or pressure.

相变图

该相变图展示了当温度或压强变化时发生的相变。

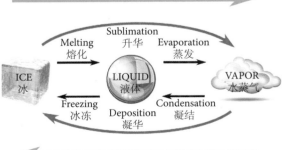

FIRST LAW of THERMODYNAMICS
热力学第一定律

TERMS	术语
isolated system	孤立系统
state of a system	系统的状态
initial state of a system	系统的初始状态
final state of a system	系统的最终状态
intermediate state of a system	系统的中间状态
microscopic state of a system	系统的微观状态
macroscopic state of a system	系统的宏观状态
cyclic process	循环过程
change in energy	能量变化
work done by a gas	气体做的功
internal energy	内能
state function	状态函数
first-law equation	第一定律方程
first-law equation for infinitesimal changes	无穷小变化的第一定律方程

LANGUAGE OF PHYSICS   物理学用语

Internal energy is the sum of the potential and kinetic energy of all the molecules of a body. A change in temperature is associated with a change in the internal energy of a gas.

内能：物体所有分子的势能和动能的总和。气体的内能发生变化时，温度也随之发生变化。

Heat is the flow of thermal energy from a body at a higher temperature to a body at a lower temperature. Heat is always positive when it is added to the system and negative when it is removed from the system.

热（量）：热能从高温物体到低温物体的流动。系统吸热时热量为正，放热时热量为负。

First law of thermodynamics
The change in the internal energy of the system equals the heat added to a system minus the work done by the system on the external environment.
This law is the law of conservation of energy as applied to a thermodynamic system.

热力学第一定律
系统的内能变化等于系统吸收的热量减去系统对外部环境做的功。
该定律是能量守恒定律在热力学系统中的应用。

A **thermodynamic system** is a system to which the equilibrium laws of thermodynamics apply.

热力学系统：热力学平衡定律所适用的系统。

WORK DONE BY A GAS

Gas contained in a cylinder at a pressure p does work on a moving piston as the system expands from a volume V to a volume $V + dV$.

(a) The work done in the expansion from the initial state to the final state is the area under the curve in a p-V diagram.

(b) The work done by a gas depends on the path between the initial and final states.

气体做的功

当系统体积从 V 增加到 $V+dV$ 时，气缸中压强为 p 的气体对运动的活塞做功。

(a) 从初状态膨胀到末状态所做的功就是 p-V 图中曲线下的面积。

(b) 气体所做的功取决于始末状态之间的路径。

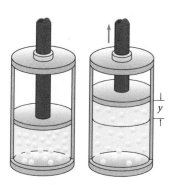

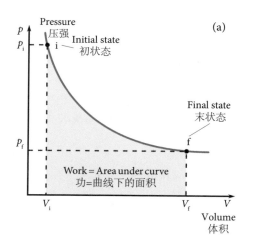

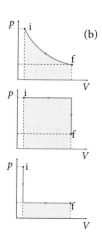

FIRST LAW OF THERMODYNAMICS

The first law of thermodynamics is a statement of the law of conservation of energy applied to a thermodynamic system.

热力学第一定律

热力学第一定律是热力学系统中的能量守恒定律。

$$\Delta U = U_f - U_i = Q - W$$

Change in energy 能量变化

Heat added to a system 系统吸收的热量

Work done by a gas 气体做的功

Internal energy at final (f) and initial (i) states
初态(i) 和末态(f)的内能

HEAT TRANSFER
热传递

TERMS	术语
surroundings	周围环境
heat conduction	热传导
law of heat conduction	热传导定律
thermal conductivity	热导率
temperature gradient	温度梯度
home insulation	房屋隔热
heat loss	热损失
convection	对流
natural convection	自然对流
forced convection	强制对流
radiation	辐射
vacuum	真空
Dewar flask	杜瓦瓶
insulating wall	隔热墙
membrane	薄膜
perfect absorber	完全吸收体
perfect emitter	完全发射体

LANGUAGE OF PHYSICS   物理学用语

Conduction is an exchange of kinetic energy between colliding molecules.

传导：分子通过碰撞交换动能的过程。

Convection is the procces whereby a heated substance moves from one place to another.

对流：高温物质从一处移动到另一处的过程。

Radiation is a transfer of thermal energy by electromagnetic waves.

辐射：利用电磁波传递热能的一种方式。

Temperature gradient is the rate at which the temperature changes with distance.

温度梯度：温度随距离的变化率。

A **conductor** is a material that transmits heat by conduction well and thus has a large value of thermal conductivity.

导体：导热性能良好、热导率高的材料。

An **insulator** is a material that is a poor conductor of heat and thus has a small value of thermal conductivity.

绝缘体：导热性能差、热导率低的材料。

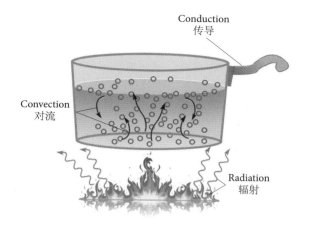

HEAT TRANSFER

Heat may be transferred by conduction, convection, and radiation.

热传递

热可以通过传导、对流和辐射来传递。

DEWAR FLASK

The thermos bottle or Dewar flask is a container designed to minimize thermal energy losses by conduction, convection, and radiation.

杜瓦瓶

保温瓶或杜瓦瓶：使传导、对流及辐射导致的热量损失最小化的容器。

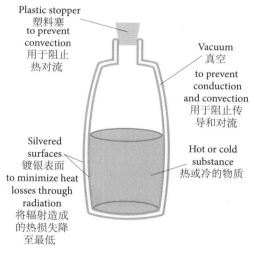

THERMOGRAM

The thermogram of a home shows colors ranging from red (areas of greatest heat loss) to blue and purple (areas of least heat loss).

热成像图

房屋热成像图呈现从红色（热量损失最大的区域）到蓝色和紫色（热量损失最小的区域）之间不同的颜色。

LAWS of RADIATION
辐射定律

TERMS 术语

Stefan-Boltzmann law	斯蒂芬-玻尔兹曼定律
Stefan-Boltzmann constant	斯蒂芬-玻尔兹曼常数
area of emitting body	发射体面积
emissivity	比辐射率/发射率
ideal absorber	理想吸收体
black body	黑体
cavity resonator	谐振腔
solar constant	太阳常数
Planck's radiation law	普朗克辐射定律
Wien displacement law	维恩位移定律

LANGUAGE OF PHYSICS 物理学用语

A **black body** is a body that absorbs all the radiation incident upon it. A black body is a perfect absorber and a perfect emitter.

黑体：吸收外来的全部辐射的物体。黑体是理想吸收体和理想发射体。

Stefan-Boltzmann law
Every body radiates energy that is proportional to the fourth power of the absolute temperature of the body.

斯蒂芬-玻尔兹曼定律
每个物体辐射的能量都与该物体的绝对温度的四次方成正比。

Planck's radiation law
An equation that shows how the energy of a radiating body is distributed over the emitted wavelengths.

普朗克辐射定律
该定律是表明物体向外辐射的能量按照波长分布情况的方程。

Wien displacement law
The product of the peak radiating wavelength and the absolute temperature is a constant.

维恩位移定律
峰值辐射波长与绝对温度的乘积是一个常数。

BLACK-BODY MODEL

A model of a black-body is the hole of a cavity in a rigid opaque body. All incoming electromagnetic radiation is absorbed and re-emitted by the rough inner surface of the cavity.

黑体模型

黑体模型：在一个不透明的刚体中有一个带孔的空腔，向孔中入射的所有电磁辐射都被腔体粗糙的内表面吸收再重新放出。

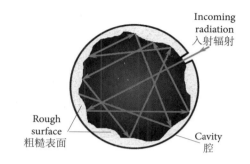

WIEN DISPLACEMENT LAW

As the temperature increases, the wavelength where the maximum or peak intensity occurs shifts to shorter wavelength.

维恩位移定律

随着温度的升高，最大或峰值强度的波长逐渐变短。

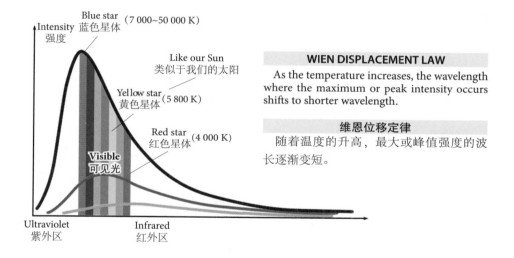

ULTRAVIOLET CATASTROPHE

The Rayleigh-Jeans law agrees with experimental results at large wavelengths (low frequencies) but strongly disagrees at short wavelengths (high frequencies). This inconsistency between observations and the predictions of classical physics is commonly known as the ultraviolet catastrophe.

紫外灾难

瑞利-金斯定律在长波（低频）范围内与实验结果较为一致，在短波（高频）范围内则与实验结果存在严重偏差。其结果和经典物理学预测之间的不一致通常被称为紫外灾难。

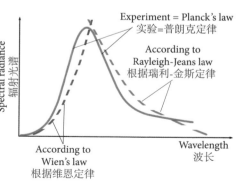

MOLECULAR MODEL of an IDEAL GAS
理想气体分子模型

TERMS	术语
ideal gas	理想气体
real gas	实际气体
monatomic ideal gas	单原子理想气体
polyatomic ideal gas	多原子理想气体
microscopic model	微观模型
short-range force	短程力
average value of the square of velocity	速度平方的平均值
root mean square speed (RMS)	均方根速率
number of molecules per unit volume	单位体积分子数

LANGUAGE OF PHYSICS 物理学用语

Assumptions of the molecular model of an ideal gas

理想气体分子模型假设

A gas is composed of a very large number of molecules that are in random motion.

气体由大量无规则运动的分子组成。

The volume of the individual molecules is very small compared to the total volume of the gas.

单个分子的体积与气体的总体积相比非常小。

The collisions of the molecules with the walls and other molecules are elastic.

分子与容器壁、分子与分子间的碰撞是弹性碰撞。

The forces between molecules are negligible except during the collision.

除碰撞瞬间外,分子间作用力可忽略。

IDEAL AND NON-IDEAL GAS BEHAVIOR

理想气体和非理想气体行为

Attractions between gas molecules serve to decrease the gas volume at constant pressure compared to an ideal gas whose molecules experience no attractive forces.

与分子间无吸引力的理想气体相比，（非理想）气体分子之间的吸引力可以在恒压下减小气体体积。

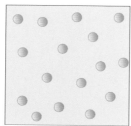

Ideal gas
理想气体

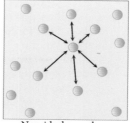

Non-ideal or real gas
非理想气体或实际气体

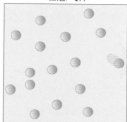

Ideal gas
理想气体

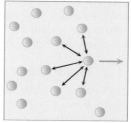

Non-ideal or real gas
非理想气体或实际气体

The attractive forces decrease the force of collisions between the molecules and container walls, therefore reducing the pressure exerted compared to an ideal gas.

相较于理想气体，吸引力减少了分子与容器壁之间的碰撞，从而减小了对容器壁的压强。

A molecule moves in the xy plane and makes an elastic collision with the wall of the container.

一个分子在xy平面上运动，并与容器壁发生弹性碰撞。

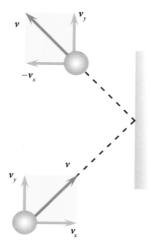

EQUIPARTITION of ENERGY
能量均分

TERMS 术语

monoatomic molecule	单原子分子
diatomic molecule	双原子分子
complex molecule	复杂分子
motion of molecule	分子的运动
translational motion of molecule	分子的平动
rotational motion of molecule	分子的转动
vibrational motion of molecule	分子的振动
degree of freedom	自由度
number of degrees of freedom	自由度的个数
equipartition theorem	均分定理
energy quantization	能量量子化

LANGUAGE OF PHYSICS 物理学用语

Degrees of freedom are the number of independent means by which a molecule can possess energy.

自由度：一个分子能独立拥有能量的方式的数目。

Theorem of equipartition of energy
The energy of a system in thermal equilibrium is equally divided among all degrees of freedom.

能量均分定理
热平衡状态下系统的能量在各个自由度之间平均分配。

Quantization of energy
Energy is quantized in some systems, meaning that the system can have only certain energies and not a continuum of energies, unlike the classical case.

能量量子化
在一些系统中，能量是量子化的，这意味着系统的能量只能取一些特定的值，而不像经典系统那样可以取一系列连续的值。

POSSIBLE MOTIONS OF A DIATOMIC MOLECULE

Translational motion of the center of mass. The center of mass of the molecule can translate in the x, y and z directions.

Rotational motion about the various axes. The molecule can rotate about three mutually perpendicular axes. We neglect the rotation about an axis for which the moment of inertia is zero.

Vibrational motion along the molecular axis:
(a) bending,
(b) symmetric stretching,
(c) asymmetric stretching.

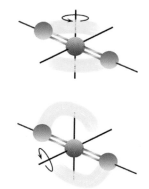
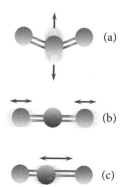

双原子分子可能的运动模式

质心的平移。分子的质心可以在x、y和z方向平移。

绕不同轴的转动。分子可以绕三个相互垂直的轴旋转。绕转动惯量为零的轴的转动忽略不计。

沿分子轴的振动：
(a)弯曲；
(b)对称拉伸；
(c)非对称拉伸。

MOLAR SPECIFIC HEAT AS A FUNCTION OF TEMPERATURE

The equipartition theorem does not explain the observed temperature variation in specific heats. It is necessary to use a quantum-mechanical model in which the energy is quantized.

摩尔比热随温度的变化

温度均分定理不能解释观测到的比热随温度的变化。有必要使用能量量子化的量子力学模型。

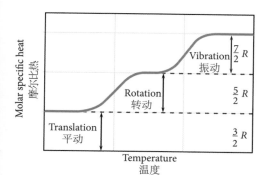

BOLTZMANN DISTRIBUTION LAW
玻尔兹曼分布律

TERMS	术语
barometric distribution	气压分布
pressure variation with altitude	压强随海拔的变化
density of the atmosphere	大气密度
distribution of molecules in the atmosphere	大气中分子的分布
weight of air	空气的重量
sea level	海平面
altitude	海拔
increasing altitude	海拔增加
the Boltzmann distribution	玻尔兹曼分布
exponential form	指数型
gravitational potential energy	重力势能
altimeter	高度仪
altimetry	测高法

LANGUAGE OF PHYSICS
物理学用语

Barometric distribution
At thermal equilibrium the density of the atmosphere decreases exponentially with increasing altitude.

气压分布
在热平衡时，大气密度随海拔的增加呈指数递减。

Boltzmann distribution law
The probability of finding the particles in a particular energy state varies exponentially as the negative of the energy divided by $k_B T$, where k_B is Boltzmann's constant and T is absolute temperature.

玻耳兹曼分布律
粒子处于特定能量状态的概率随能量除以 $k_B T$ 的负值呈指数型变化，其中 k_B 是玻尔兹曼常数，T 是绝对温度。

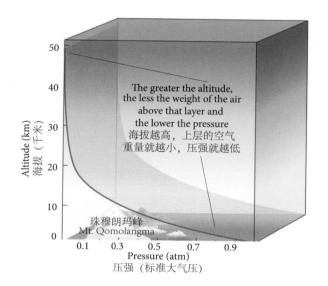

LAW OF ATMOSPHERES

The pressure in the atmosphere decreases as the altitude increases.

大气定律

大气压强随海拔的升高而降低。

An altimeter is an instrument used to measure the altitude of an object above a fixed level. Altitude can be determined based on the measurement of atmospheric pressure. An altimeter is found in most aircrafts.

高度仪是用来测量相对于某固定平面的高度的仪器。高度可以通过测量大气压来计算得到。大多数飞机上都装有高度仪。

BOLTZMANN DISTRIBUTION LAW

All particles would fall into the lowest energy level, except that the thermal energy $k_B T$ tends to excite the particles to higher energy levels.

玻耳兹曼分布律

所有粒子都将落入最低能级，除非热能$k_B T$能够将粒子激发到更高能级。

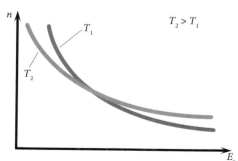

DISTRIBUTION of MOLECULAR SPEED
分子速率分布

TERMS 术语

speed distribution	速率分布
function of speed distribution	速率分布函数
curve of speed distribution	速率分布曲线
speed distribution curve reaches a peak	速率分布曲线达到峰值
RMS speed	均方根速率/方均根速率
average speed	平均速率
most probable speed	最概然速率
[Maxwell-]Boltzmann distribution function	[麦克斯韦-]玻尔兹曼分布函数
range from v to $v+dv$	在v到$v+dv$区间
molecules with speeds between v and $v+dv$	速率在v到$v+dv$区间的分子
probability	概率
asymmetric shape	不对称的形状

LANGUAGE OF PHYSICS 物理学用语

[Maxwell-] Boltzmann distribution function [麦克斯韦-]玻尔兹曼分布函数
The fundamental expression that describes the most probable distribution of speeds of N gas molecules in terms of the mass of a gas molecule and the absolute temperature.

用单个气体分子质量和绝对温度来表示N个气体分子最概然速率分布的基本表达式。

The **most probable speed** is the speed at which the distribution curve reaches a peak.

最概然速率：分布曲线达到峰值时的速率。

The **root mean square (RMS) speed** of the molecules is the square root of the average value of the square of the velocity for all the molecules.

分子的**均方根速率**：所有分子速度平方平均值的平方根。

SPEED DISTRIBUTION

The speed distribution of gas molecules at some temperature. The number of molecules in the range dv is equal to the area of the shaded rectangle $N_v dv$. The function N_v approaches zero as v approaches infinity.

速率分布

特定温度下气体分子的速率分布。在 dv 范围内的分子数等于阴影矩形的面积 $N_v dv$。当 v 趋于无穷时,函数 N 趋于零。

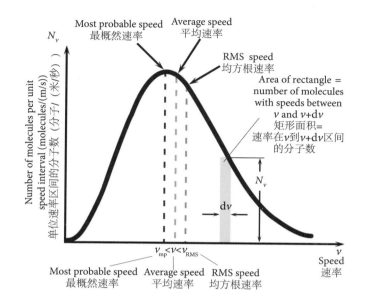

SPEED DISTRIBUTION FUNCTION AT DIFFERENT TEMPERATURES

The curve shifts to the right and broadens as T increases. The area under each of the three curves is the same.

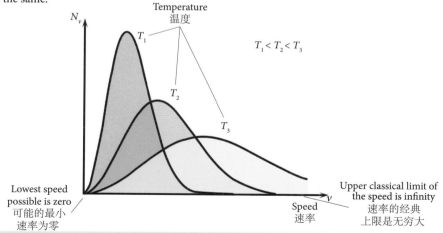

不同温度下的速率分布函数

随着 T 的增加,曲线变宽并向右移动。三条曲线下的面积相等。

MEAN FREE PATH
平均自由程

TERMS	术语
mean free path	平均自由程
random-walk process	随机游走过程
mean distance	平均距离
number of collisions	碰撞次数
average distance	平均距离
collision frequency	碰撞频率
mean free time	平均自由时间
effective diameter	有效直径
point mass	质点
spherical molecules collision	球形分子碰撞
equivalent collision	等效碰撞

LANGUAGE OF PHYSICS   物理学用语

The **mean free path** is the average distance between collisions.

平均自由程：碰撞之间的平均距离。

Collision frequency is the number of collisions per unit time.

碰撞频率：单位时间内的碰撞次数。

The **mean free time** is the average time between collisions or the inverse of the collision frequency.

平均自由时间：碰撞之间的平均时间或碰撞频率的倒数。

RANDOM MOTION

A molecule moving through a gas collides with other molecules in a random fashion. This behavior is referred to as a random-walk process.

随机运动

在气体中运动的分子随机地与其他分子碰撞。这种行为被称为随机游走过程。

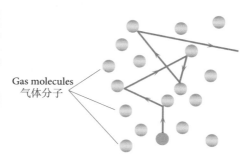

Gas molecules
气体分子

EFFECTIVE DIAMETER

(a) Two spherical molecules collide if their centers are within a distance d of each other.
(b) The collision between the two molecules is equivalent to a point mass colliding with a molecule having an effective diameter of $2d$.

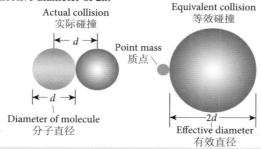

Actual collision
实际碰撞

Equivalent collision
等效碰撞

Point mass
质点

Diameter of molecule
分子直径

Effective diameter
有效直径

有效直径

(a)两个球形分子的中心相距 d 以内时发生碰撞。
(b)这两个分子间的碰撞等效于一个质点与一个有效直径为 $2d$ 的分子的碰撞。

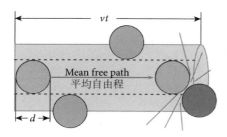

Mean free path
平均自由程

MEAN FREE PATH

A molecule of effective diameter $2d$ collides with every molecule within cylinder of length vt, where v is its average speed.

平均自由程

一个有效直径为 $2d$ 的分子与长度为 vt 的圆柱体内的每个分子发生碰撞,其中 v 为其平均速度。

VAN DER WAALS' EQUATION
范德华方程

TERMS 术语

ideal gas	理想气体
real gas	真实气体
liquefied gas	液化气体
assumption	假设
negligible intermolecular force	可忽略的分子间作用力
negligible volume occupied by molecule	可忽略的分子体积
modified equation of state	改良状态方程
empirical constant	经验常数
hyperbolic curve	双曲线
approximately hyperbolic curve	近似双曲线
nonlinear curve	非线性曲线
experimental curve	实验曲线
isotherm	等温线
critical pressure p_c	临界压强 p_c
critical temperature T_c	临界温度 T_c
critical volume V_c	临界体积 V_c
liquid state	液态
gaseous state	气态
liquid-vapor equilibrium state	气液平衡态

LANGUAGE OF PHYSICS 物理学用语

The **van der Waals' equation of state** is a modification of the ideal gas law which takes into account molecular size and molecular interaction forces.

范德华状态方程：理想气体定律的一种改进，它的特点在于考虑了分子的大小以及分子之间的相互作用力。

The **critical temperature** of a substance is the temperature above which the vapor of the substance cannot be liquefied, no matter how much pressure is applied.

临界温度：高于此温度时，无论施加多大的压强，物质都不能从气态转化为液态。

The **critical pressure** of a substance is the pressure required to liquefy a gas at its critical temperature.

临界压强：临界温度时物质液化所需要的压强。

The **critical volume** is the volume of a unit mass (usually one mole) of a substance measured when it is at its critical temperature and pressure.

临界体积：临界温度、临界压强条件下，测得的物质单位质量的体积（通常为1摩尔）。

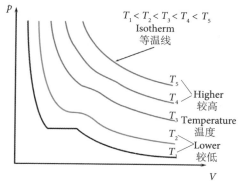

THEORETICAL ISOTHERMS

Isotherms for a real gas at various temperatures. At a higher temperature, such as T_5, the behavior is nearly ideal. The behavior is not ideal at the lower temperature.

理论等温线

不同温度下真实气体的等温线。在较高温度，如T_5时，真实气体的表现接近于理想气体；在较低温度时，真实气体的表现则与理想气体存在较大偏差。

VAN DER WAALS' EQUATION

Equation of state for a real gas.

范德华方程

真实气体的状态方程。

$$\left(p + \frac{n^2 a}{V^2}\right)\left(\frac{V}{n} - b\right) = RT$$

EXPERIMENTAL ISOTHERMS

The experimental curves for CO_2 are described quite accurately by van der Waals' equation at the higher temperatures (T_3, T_4, and T_5) and outside the shaded region. Below the critical temperature T_c, the substance can occur in the liquid state, the liquid-vapor equilibrium state, or the gaseous state, depending on the pressure and volume.

实验等温线

范德华方程可以十分准确地描述CO_2在较高温度（T_3、T_4和T_5）以及在阴影部分以外的实验曲线。临界温度T_c以下时，物质可能处在液态、气液平衡态或者气态，具体情况取决于压强和体积。

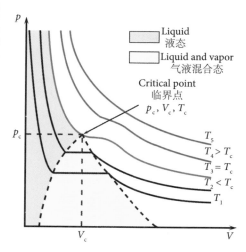

SECOND LAW of THERMODYNAMICS
热力学第二定律

TERMS 术语	
process	过程
reversible process	可逆过程
irreversible process	不可逆过程
cyclic process	循环过程
cycle	循环
thermal energy	热能
thermal energy absorption	热能吸收
net work done	做的净功
the second law of thermodynamics	热力学第二定律
Kelvin-Planck's statement of the second law	第二定律的开尔文-普朗克表述
Clausius' statement of the second law	第二定律的克劳修斯表述
heat engine	热机
hot reservoir	高温热源
cold reservoir	低温热源
working substance	工作介质

LANGUAGE OF PHYSICS　　　　　　　物理学用语

Kelvin-Planck's statement of the second law
It is impossible to construct a heat engine that, operating in a cycle, produces no effect other than the conversion of thermal energy from a reservoir into an equal amount of mechanical work.

第二定律的开尔文-普朗克表述
在一次循环中，能够从单一热源吸收热能并转化为等量机械功而不产生其他影响的热机是无法制造出来的。

Clausius' statement of the second law
It is impossible to construct a cyclical machine whose sole effect is to transfer heat continuously from one body to another body at a higher temperature.

第二定律的克劳修斯表述
唯一作用是将物体的热量源源不断地转移到高温物体上去的循环运作的热机是无法制造出来的。

IMPOSSIBLE HEAT FLOW

Thermal energy does not flow spontaneously from a cold object to a hot object. When two objects at different temperatures are placed in thermal contact with each other, thermal energy always flows from the warmer to the cooler object.

不可能的热流

热能不会自发地从低温物体流向高温物体。两个温度不同的物体发生热接触时，热能总是从高温物体流向低温物体。

"PERFECT" HEAT ENGINE

Schematic diagram of the impossible "perfect" heat engine. It is impossible to construct such a perfect engine: a heat engine that absorbs thermal energy Q_H from a hot reservoir and does an equivalent amount of work W.

"完美"热机

不存在的"完美"热机示意图。能够从热源吸收热能 Q_H 并转换为等量的功 W 的"完美"热机是制造不出来的。

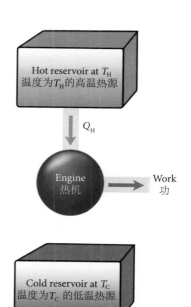

HEAT ENGINE and REFRIGERATOR
热机和制冷机

TERMS 术语	
heat engine	热机
heat pump	热泵
refrigerator	制冷机
thermal efficiency	热效率
coefficient of performance (COP)	制热系数
thermal energy	热能
absorbed thermal energy	吸收的热能
expelled thermal energy	释放的热能

LANGUAGE OF PHYSICS 物理学用语

A **heat engine** is a device that converts thermal energy to other forms of energy.

热机：一种能够将热能转化为其他能量形式的装置。

A **refrigerator** and a heat pump are heat engines running in reverse.

制冷机和热泵是作用相反的两种热机。

The **thermal efficiency of a heat engine** is the ratio of the net work done to the thermal energy absorbed at the higher temperature over one complete cycle.

热机的热效率：在一个完整循环中，对外做的净功与从高温热源吸收的热能之比。

The **coefficient of performance** is the ratio of the heat transferred into the hot reservoir and the work required to transfer that heat.

制热系数：向高温热源输入的热量与输送热量所做的功之比。

热力学 **145**

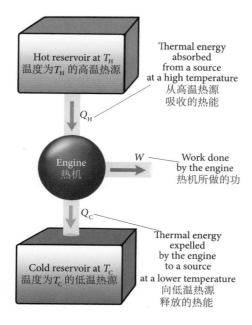

SCHEMATIC DIAGRAM OF A HEAT ENGINE

The engine (1) absorbs thermal energy Q_H from the hot reservoir, (2) expels thermal energy Q_C to the cold reservoir, and (3) does work W.

热机示意图

热机（1）从高温热源吸收热能Q_H，（2）向低温热源释放热能Q_C，（3）对外做功W。

SCHEMATIC DIAGRAM OF A REFRIGERATOR

The engine (1) absorbs thermal energy Q_C from the cold reservoir, (2) expels thermal energy Q_H to the hot reservoir, and (3) work W is done on the refrigerator.

制冷机示意图

制冷机（1）从低温热源吸收热能Q_C，（2）向高温热源释放热量Q_H，（3）对制冷机做功W。

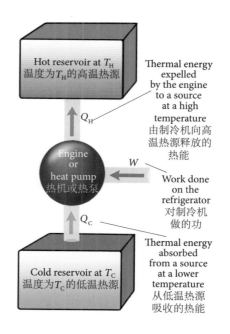

CARNOT ENGINE
卡诺热机

TERMS 术语

engine	发动机
heat engine	热机
theoretical engine	理论发动机
most efficient engine	最高效热机
Carnot cycle	卡诺循环
Carnot engine	卡诺热机
Carnot's theorem	卡诺定理
adiabatic process	绝热过程
isothermal process	等温过程
reversible process	可逆过程
expansion	膨胀
compression	压缩
efficiency of a Carnot engine	卡诺热机效率
heat loss	热损失
heat loss by conduction	传导热损失
heat loss by friction	摩擦热损失

LANGUAGE OF PHYSICS 物理学用语

Carnot's theorem
No real heat engine operating between two heat reservoirs can be more efficient than a Carnot engine operating between the same two reservoirs.

卡诺定理
在两个热源间工作的实际热机不可能比在两个相同热源间工作的卡诺热机有更高的（热）效率。

The **Carnot cycle** consists of two adiabatic and two isothermal processes, all reversible.

卡诺循环：由两个绝热过程和两个等温过程构成，并且它们都是可逆的。

THE CARNOT CYCLE

The p-V diagram for the Carnot cycle. Points A, B, C, and D are points on a P-V diagram.

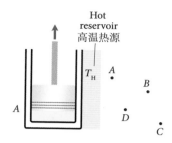

卡诺循环

卡诺循环的压强-体积图（p-V图）。点 A、B、C和D都是 p-V 图上的点。

In process $A \to B$, the gas expands isothermally while in contact with a reservoir at T_H.

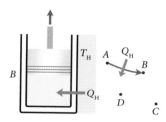

在 $A \to B$ 的过程中，气体与温度为 T_H 的热源接触，进行等温膨胀。

In process $B \to C$, the gas expands adiabatically ($Q=0$).

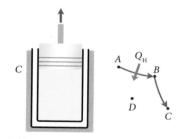

在 $B \to C$ 过程中，气体进行绝热膨胀($Q=0$)。

In process $C \to D$, the gas is compressed isothermally while in contact with a reservoir at $T_C < T_H$.

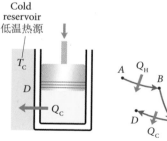

在 $C \to D$ 过程中，气体与温度为 T_C（$T_C < T_H$）的热源接触，被等温压缩。

In process $D \to A$, the gas is compressed adiabatically.

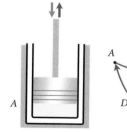

在 $D \to A$ 过程中，气体被绝热压缩。

GASOLINE ENGINE
汽油发动机

TERMS 术语

engine	发动机
gasoline engine	汽油发动机
conventional engine	传统发动机
diesel engine	柴油发动机
Stirling engine	斯特林发动机
steam engine	蒸汽机
cycle	循环
Otto cycle	奥托循环
Carnot cycle	卡诺循环
four-stroke cycle	四冲程循环
intake stroke	进气冲程
intake valve	进气阀
compression stroke	压缩冲程
combustion	燃烧
power stroke	做功冲程
ignition	点火
exhaust stroke	排气冲程
piston	活塞
spark plug	火花塞
air-fuel mixture	油气混合物
adiabatically compressed	绝热压缩
adiabatically expanded	绝热膨胀
temperature increases	升温
temperature rises rapidly	快速升温
temperature drops from T_1 to T_2	温度从 T_1 降到 T_2
work done by a gas	气体做的功
area under a curve	曲线下的面积
exhaust valve	排气阀
residual gases	剩余气体
efficiency of cycle	循环效率
atmospheric pressure	大气压

LANGUAGE OF PHYSICS 物理学用语

An **engine** is a machine designed to convert other forms of energy into mechanical energy.

发动机：将其他形式的能量转化为机械能的一种机器。

A **heat engine** is a system that converts heat or thermal energy and chemical energy into mechanical energy, which can then be used to do mechanical work.

热机：一个将热能和化学能转化为机械能以用于做机械功的系统。

热力学 149

(a) Intake
吸气

(b) Compression
压缩

GASOLINE ENGINE

The four-stroke cycle of a conventional gasoline engine.

(a) In the intake stroke, air is mixed with fuel.
(b) The intake valve is then closed, and the air-fuel mixture is compressed by the piston.
(c) The mixture is ignited by the spark plug raising it to a higher temperature. In the power stroke, the gas expands against the piston.
(d) In the exhaust stroke the spent air-fuel mixture expels through the exhaust valve. After that the cycle repeats.

汽油发动机

传统汽油发动机的四冲程循环：

（a）在吸气冲程中，空气与燃料混合。

（b）吸气阀关闭，空气-燃料混合物被活塞压缩。

（c）混合物被火花塞点燃，温度升高。在做功冲程中，气体膨胀，压缩活塞。

（d）在排气冲程中，燃烧过的空气-燃料混合物从排气阀中释放。然后整个循环重复。

(c) Power
做功

(d) Exhaust
排气

OTTO CYCLE

p-V diagram of the Otto cycle.
O-A is intake stroke
A-B is compression stroke
B-C is combustion
C-D is power stroke
A-O is exhaust stroke

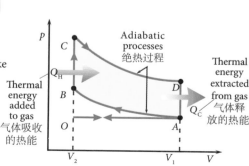

奥托循环

奥托循环的压力-体积图（p-V图）。
$O \rightarrow A$：吸气冲程
$A \rightarrow B$：压缩冲程
$B \rightarrow C$：燃烧
$C \rightarrow D$：做功冲程
$A \rightarrow O$：排气冲程

ENTROPY
熵

TERMS	术语
entropy function	熵函数
entropy function increases	熵函数增加
entropy function decreases	熵函数减少
state function	状态函数
equilibrium state	平衡状态
initial equilibrium state	初始平衡状态
final equilibrium state	最终平衡状态
infinitesimal process	无限小过程
isentropic process	等熵过程
reversible process	可逆过程
irreversible (real) process	不可逆（实际）过程
amount of thermal energy	热能量
absolute temperature	绝对温度
change in entropy	熵变
isolated system	孤立系统
disorder	无序
heat death of the Universe	热寂
probability	概率

LANGUAGE OF PHYSICS 物理学用语

Entropy is a measure of the disorder in a system.

熵：衡量体系混乱程度的量度。

The change in entropy for any reversible, cyclic process is zero.

任何可逆、循环的过程的熵都不变。

In an irreversible process, the total entropy of an isolated system always increases.

不可逆过程中，孤立系统总的熵总是增加的。

The entropy of the University is unknown, but is likely to remain constant or increase.

宇宙的熵未知，但很可能为一个常量或持续增加。

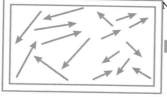
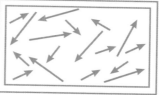

(a) The container of gas has molecules with speeds above the mean value on the left side, and molecules with lower than the mean value on the right side.
(b) The random motions tend to mix the slow- and fast-moving molecules uniformly. Because of the statistical tendency of systems to proceed toward states of greater probability and greater disorder, all natural processes are irreversible and entropy increases.

（a）左边容器中气体分子的运动速率高于平均值，而右边容器中气体分子的速率低于平均值。
（b）随机运动将慢速和快速运动的分子均匀混合在一起。从统计学上看，任何系统都趋向于概率更大和更无序的状态，因此所有的自然过程都是不可逆且熵增的过程。

Demonstrating the second law of thermodynamics.
The only changes that are possible for an isolated system are those in which the entropy of the system either increases or remains the same. The broken plate will not pull itself back together.

热力学第二定律的演示。
对于一个孤立系统来说，唯一可能发生的变化是系统的熵增加或者保持不变。比如，摔碎的盘子无法自行复原。

ELECTROMAGNETISM

A team of scientists has discovered the first robust example of a new type of magnet USb_2 that holds promise for enhancing the performance of data storage technologies. This singlet-based magnet differs from conventional magnets, in which small magnetic constituents align with one another to create a strong magnetic field. By contrast, the newly uncovered singlet-based magnet has fields that pop in and out of existence, resulting in an unstable force—but also one that potentially has more flexibility than conventional counterparts.

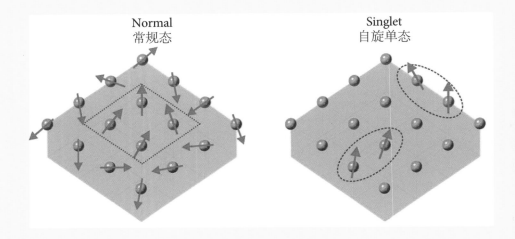

科学家们首次发现了有望提高数据存储技术性能的新型磁体USb_2。这种基于自旋单态的磁体不同于传统的磁体。传统的磁体中，小的磁性成分彼此平行，形成一个强大的磁场。相比之下，新发现的基于自旋单态的磁体的磁场会突然出现或消失，从而产生一种不稳定的力，但这种力也可能比传统磁体更具灵活性。

电磁学

ELECROMAGNETISM

Electromagnetism is a branch of physics involving the study of electrical charge and of the forces and fields associated with charge.

Electricity and **magnetism** are two aspects of electromagnetism.

BRANCHES OF ELECTROMAGNETISM

Electricity is the set of physical phenomena associated with the presence of electric charges.

Magnetism is a class of physical phenomena associated with magnetic fields, which arise from the motion of electric charges.

电磁学

电磁学：物理学的一个分支，研究电荷以及与电荷相关的力和场。

电和磁是电磁学研究的两个方面。

电磁学分支

电：与电荷的存在有关的一系列物理现象。

磁：与磁场有关的一类物理现象，它是由电荷的运动而产生的。

ELECTRIC CHARGE
电荷

TERMS	术语
electrostatics	静电学
electron	电子
amber	琥珀
rubber or glass rod	橡胶或玻璃棒
rubbed with silk or fur	用丝绸或毛皮摩擦
electrified	通电的
electrically charged	带电的
charge	电荷
positive charge	正电荷
negative charge	负电荷
like charges	同性电荷
unlike charges	异性电荷
induced charge	感应电荷
charge conserved	电荷守恒的
quantized charge	量子化的电荷
repulsive force	排斥力
attractive force	吸引力
conductor	导体
grounded conductor	接地导体
insulator	绝缘体
semiconductor	半导体
induction	电磁感应
conduction	（电）传导
electroscope	验电器
electrometer	静电计

LANGUAGE OF PHYSICS   物理学用语

Electrostatics is a branch of physics that studies electric charges at rest under the action of electric forces.

静电学：物理学的一个分支，研究在电力作用下静止的电荷。

Electrostatic charge is the term given to charge that is stationary i.e. not flowing in the form of a current.

静电电荷：静止的电荷，即不以电流形式流动的电荷。

Unlike charges attract one another and like charges repel one another.

异性电荷相互吸引，同性电荷相互排斥。

The law of conservation of electric charge
Electric charge can neither be created nor destroyed.

电荷守恒定律
电荷既不能被创造，也不能被消灭。

POSITIVE AND NEGATIVE CHARGES

There are two kinds of charges in nature, with the property that unlike charges attract one another and like charges repel one another.

正电荷与负电荷

自然界中有两种电荷，异性电荷相互吸引，同性电荷相互排斥。

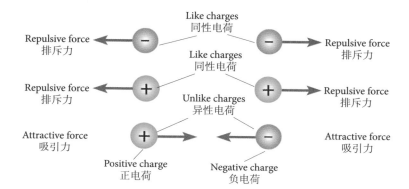

Heliopause Electrostatic Rapid Transit System (HERTS) is a revolutionary propellant-less propulsion concept that is ideal for deep space missions to the outer planets. The basic principle on which the HERTS operates is the exchange of momentum between an array of long electrically biased wires that extend outward 10 to 30 km from a rotating spacecraft and the protons from the solar wind which flow radially away from the Sun at speeds ranging from 300 to 700 km/s. The spacecraft can be accelerated to enormous speeds of 100-150 km/s (~ 20 to 30 AU/year).

日球层顶静电快速传输系统（HERTS）是一种革命性的无推进剂推进概念，是前往外行星完成深空任务的理想选择。HERTS运行的基本原理是从旋转的航天器向外延伸出一组10~30千米长的带电导线，和太阳风中的质子发生动量交换。太阳风是从太阳中放射出来的，速度可以达到300~700千米/秒。宇宙飞船可以加速至100~150千米/秒（约20~30个天文单位/年）。

COULOMB'S LAW
库仑定律

TERMS 术语

fundamental law	基本定律
particle	粒子
stationary particle	静止粒子
charged particle	带电粒子
point charge	点电荷
point charges of opposite sign	符号相反的点电荷
point charges of same sign	符号相同的点电荷
magnitude of charge	电荷量
electrostatic force	静电力
electric force	（静）电力
magnitude of the electric force	电力的大小
attractive electric force	电吸引力
repulsive electric force	电排斥力
proportional	成正比的
inversely proportional	成反比的
directed along the line	沿直线方向
distance of separation	间隔距离
square of the separation	间距的平方
Coulomb constant	库仑常数
permittivity of free space	真空介电常数
force caused by multiple charges	由多个电荷引起的力

LANGUAGE OF PHYSICS 物理学用语

Coulomb's law
The electrostatic force between two stationary, charged particles separated by a distance r is proportional to the product of the charges q_1 and q_2 on the two particles, and inversely proportional to the square of the separation and directed along the line joining the particles.

库仑定律
距离为 r 的两个静止的带电粒子之间的静电力，与两个粒子所带电荷 q_1 和 q_2 的乘积成正比，与粒子间距离的平方成反比，且力的方向与两个粒子的连线一致。

Force caused by multiple charges
If there are three or more charges present, then the force on any one charge is found by the vector addition of the forces associated with the other charges.

由多个电荷引起的力
如果存在三个或三个以上的电荷，那么其中任意一个电荷所受到的力，等于其与其他电荷之间作用力的矢量叠加。

FORCES BETWEEN ELECTRIC CHARGES

Two point charges separated by a distance r exert a force on each other given by Coulomb's law.

(a) When the charges are of the same sign, the force is repulsive.

(b) When the charges are of the opposite sign, the force is attractive.

电荷间的力

根据库仑定律，相距r的两个点电荷产生相互作用的力。

（a）电荷符号相同时，产生排斥力。

（b）电荷符号相反时，产生吸引力。

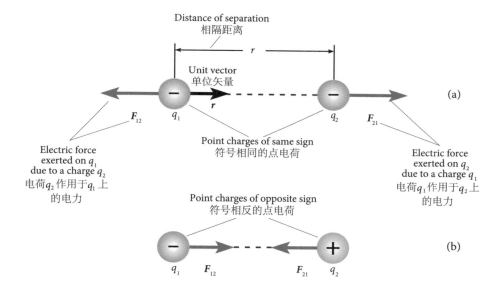

COULOMB'S LAW

Coulomb's law states the magnitude of the electric force between two charges.

库仑定律

库仑定律给出了两个电荷之间电力的大小。

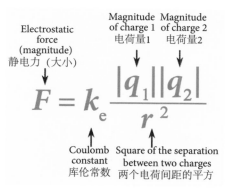

ELECTRIC FIELD
电场

TERMS	术语
electric force	电力
charge	电荷
point charge	点电荷
external charge	外部电荷
charge distribution	电荷分布
continuous charge distribution	连续电荷分布
charge density	电荷密度
volume charge density	体电荷密度
surface charge density	面电荷密度
linear charge density	线电荷密度
test charge	检验电荷
positive test charge	正检验电荷
test charge magnitude	检验电荷带电量
test charge at rest	静止的检验电荷
electric field vector	电场矢量
electric field line	电场线
electric field strength	电场强度
directed radially outward	沿径向向外
directed towards the charge	指向电荷
superposition principle	叠加原理
a group of charges	一组电荷

LANGUAGE OF PHYSICS
物理学用语

An **electric field** exists in the space around an electric charge. It is an intrinsic property of nature.

电场存在于电荷周围的空间中。它是自然的固有属性。

The **electric field** E at some point in space is defined as the electric force F that acts on a small positive test charge placed at that point divided by the magnitude of the test charge.

空间中某一点上的**电场** E：作用于放置在该点上的小的正检验电荷的电场 F 除以检验电荷的电量。

The **superposition principle of electric fields** states that the total electric field at a point due to a group of charges equals the vector sum of the electric fields of all the charges at that point.

电场的叠加原理表明，某一点上由一组电荷引起的总电场等于所有电荷在该点电场的矢量和。

ELECTRIC FIELD DIRECTION

A positive test charge q_0 at point B is a distance r from a point charge q. The electric field at B points (a) radially inward toward q if q is negative, (b) radially outward from q if q is positive.

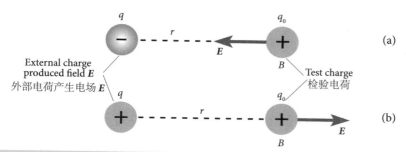

External charge produced field E
外部电荷产生电场 E

Test charge
检验电荷

电场方向

在 B 点的正检验电荷 q_0 与点电荷 q 的距离为 r。B 点的电场：(a) 当 q 为负时，沿径向指向 q；(b) 当 q 为正时，从 q 径向向外。

POSITIVE CHARGE

The electric field lines for a positive point charge.

正电荷

正点电荷的电场线。

NEGATIVE CHARGE

The electric field lines for a negative point charge.

负电荷

负点电荷的电场线。

UNLIKE POINT CHARGES

The electric field lines for two equal and opposite point charges.

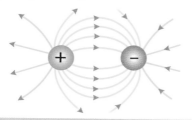

异性点电荷

两个带电量相等、符号相反的点电荷的电场线。

LIKE POINT CHARGES

The electric field lines for two positive point charges.

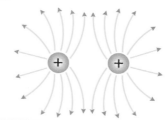

同性点电荷

两个正点电荷的电场线。

ELECTRIC FLUX
电通量

TERMS 术语

electric field	电场
uniform electric field	均匀电场
uniform electric field in magnitude	大小均匀的电场
uniform electric field in direction	方向均匀的电场
electric field line	电场线
electric field strength	电场强度
number of lines per unit area	单位面积的线数
number of lines penetrating surface	穿过曲面的线数
normal to surface	垂直于曲面
projected area	投影面积
electric flux	电通量
net electric flux	净电通量
electric flux through surface	通过曲面的电通量
electric flux leaving surface	离开曲面的电通量
electric flux entering surface	进入曲面的电通量
positive electric flux	正电通量
negative electric flux	负电通量
plane	平面
surface of area A	面积为A的曲面
closed surface	封闭曲面
integral over a closed surface	封闭曲面的积分

LANGUAGE OF PHYSICS 物理学用语

Electric flux is represented by the number of electric field lines penetrating some surface.

电通量用穿过某一表面的电场线的数目来表示。

Electric flux is the product of the electric field intensity E and a surface area A perpendicular to the field.

电通量：电场强度 E 和垂直于电场的表面面积 A 的乘积。

The **net flux through a closed surface** can be found as an integral over a closed surface of the product of the component of the electric field normal to the surface E_n and area element dA.

可以通过对电场的垂直分量E_n与面积元dA的乘积在封闭曲面上做积分求得**通过封闭曲面的净通量**。

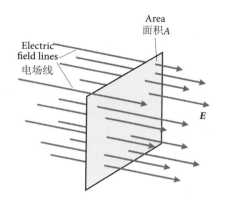

PLANE PERPENDICULAR TO FIELD

Uniform electric field lines penetrating a plane of area A perpendicular to the field. The electric flux through this perpendicular area is equal to EA.

垂直于场的平面

穿过面积为A、垂直于电场的平面的均匀电场线。通过该垂直区域的电通量等于EA。

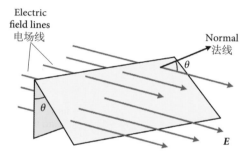

PLANE AT ANGLE TO FIELD

Uniform electric field lines through an area A whose normal is at angle θ to the field. The electric flux through this surface is equal to $EA\cos\theta$.

与场成一定角度的平面

通过面积为A、法线与电场夹角为θ的平面的均匀电场线。通过该表面的电通量等于$EA\cos\theta$。

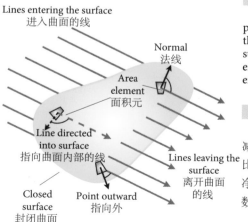

CLOSED SURFACE

The net flux through a closed surface is proportional to the net number of lines leaving the surface minus the number entering the surface. If there are more lines leaving than entering, the net flux is positive. If more lines enter than leave, the net flux is negative.

封闭曲面

通过封闭曲面的净通量与离开曲面的线数减去进入曲面的线数后得到的净剩值成正比。若离开曲面的线数多于进入的线数，则净通量为正；若进入的线数多于离开的线数，则净通量为负。

GAUSS' LAW
高斯定律

TERMS 术语

net electric flux	净电通量
net electric flux through a closed surface	通过封闭曲面的净电通量
Gaussian surface	高斯面
spherical Gaussian surface	球体高斯面
nonspherical Gaussian surface	非球体高斯面
closed Gaussian surface	封闭高斯面
boundary	边界
charge enclosed by surface	曲面覆盖的电荷
charge inside surface	曲面内电荷
charge distribution	电荷分布
symmetric charge distribution	对称电荷分布
planar symmetry	平面对称
cylindrical symmetry	柱对称
spherical symmetry	球对称
net charge	净电荷
permittivity of free space	真空介电常数

LANGUAGE OF PHYSICS　　　　物理学用语

Gauss' law
The net electric flux through any closed Gaussian surface is equal to the net charge inside the surface divided by the dielectric constant.

高斯定律
通过任意封闭高斯面的净电通量等于该曲面内的净电荷除以介电常数。

Gauss' law is useful for evaluating the electric field of charge distributions that have spherical, cylindrical, or planar symmetry.

高斯定律可用于计算具有球对称、柱对称或平面对称的电荷分布的电场。

ELECTRIC FLUX THROUGH A CLOSED SURFACE

The net electric flux through any closed surface depends only on the charge inside that surface.

通过封闭曲面的电通量

通过任意封闭曲面的净电通量只取决于该曲面内的电荷。

- Electric charge 电荷
- Closed surfaces of various shapes 各种形状的封闭曲面
- Electric field lines 电场线

GAUSS' LAW

Using Gauss' law, one can calculate the electric field due to various symmetric charge distributions.

高斯定律

利用高斯定律可以计算各种具有对称性的电荷分布所引起的电场。

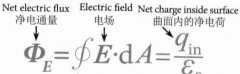

$$\Phi_E = \oint \vec{E} \cdot d\vec{A} = \frac{q_{in}}{\varepsilon_0}$$

- Net electric flux 净电通量
- Electric field 电场
- Net charge inside surface 曲面内的净电荷
- Integral over a closed surface 封闭曲面的积分
- Differential vector area 微分向量面积
- Permittivity of free space 真空介电常数

GAUSSIAN AND NON-GAUSSIAN SURFACES

The surface of a sphere, surface of a torus, and surface of a cube are some Gaussian surfaces. The surface of a disk, a square surface, or a hemispherical surface cannot be used as Gaussian surfaces, as they contain boundaries.

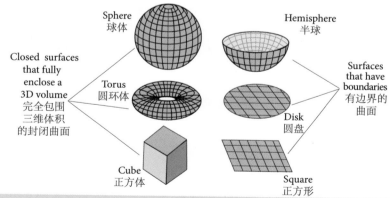

- Closed surfaces that fully enclose a 3D volume 完全包围三维体积的封闭曲面
- Sphere 球体
- Torus 圆环体
- Cube 正方体
- Hemisphere 半球
- Disk 圆盘
- Square 正方形
- Surfaces that have boundaries 有边界的曲面

高斯面和非高斯面

球体、圆环体和正方体的表面都是高斯面。圆盘、正方形或半球的表面不能用作高斯面，因为它们都有边界。

ELECTROSTATIC EQUILIBRIUM
静电平衡

TERMS 术语	
conductor	导体
isolated conductor	孤立导体
charged conductor	带电导体
irregularly shaped conductor	形状不规则的导体
conductor in electrostatic equilibrium	静电平衡的导体
free electron	自由电子
motion of charges	电荷运动
electrostatic generator	静电起电机
van de Graaff generator	范德格拉夫起电机
electrostatic shielding or Faraday cage	静电屏蔽或法拉第笼

LANGUAGE OF PHYSICS 物理学用语

A **conductor** is a material that permits the free flow of electric charge through it.

导体：一种允许电荷自由通过的材料。

Properties of a conductor in electrostatic equilibrium

静电平衡时导体的性质

1. The electric field is zero everywhere inside the conductor.

1. 电场在导体内部处处为零。

2. Any charge on an isolated conductor resides on its surface.

2. 电荷分布在孤立导体的表面。

3. The electric field just outside a charged conductor is perpendicular to its surface.

3. 带电导体外部的电场垂直于其表面。

4. On an irregularly shaped conductor, charge tends to accumulate at sharp points.

4. 在形状不规则的导体上，电荷往往聚集在导体尖端。

A CONDUCTOR IN AN EXTERNAL ELECTRIC FIELD

The charges induced on the surface of the conductor produce an electric field that opposes the external field, giving a resultant field of zero inside the conductor.

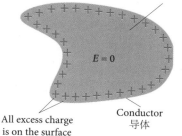

Electric field inside the conductor is zero
导体内部的电场为零

$E = 0$

All excess charge is on the surface
所有多余的电荷都在表面

Conductor
导体

外电场中的导体

在导体表面的感应电荷产生一个与外场相反的电场，使得导体内部总电场为零。

FARADAY CAGE

Faraday cage (shield) at a power plant. A Faraday cage is an enclosure used in order to block external electric fields.

法拉第笼

发电厂的法拉第笼（屏障）。法拉第笼是一种用来阻挡外部电场的封闭装置。

VAN DE GRAAFF GENERATOR

A van de Graaff generator is an electrostatic generator which accumulates electric charges on a hollow metal globe, creating very high electric potentials. The potential difference achieved in modern van de Graaff generators can reach 5 megavolts.

范德格拉夫起电机

范德格拉夫起电机是一种静电发生器，它在空心金属球上积聚电荷，由此产生非常高的电势。现代范德格拉夫起电机的电位差可以达到5兆伏。

ELECTRIC POTENTIAL
电势

TERMS	术语
potential	（电）势
potential of a point charge	点电荷的电势
potential of a charged conductor	带电导体的电势
potential energy	势能
potential of a pair of point charges	一对点电荷的电势
superposition of potentials	势的叠加
potential difference (voltage)	电势差（电压）
equipotential line	等势线
equipotential surface	等势面
path integral or line integral	路径积分或线积分
independent of the path	与路径无关
partial derivative	偏导数
gradient operator	梯度算符
field-ion microscope	场离子显微镜
field-ion microscope image	场离子显微镜图像

LANGUAGE OF PHYSICS   物理学用语

Electric potential is the potential energy per unit charge.

电势：单位电荷的势能。

Potential is a scalar quantity.

势是一种标量。

Superposition of potentials states that the total potential at any arbitrary point due to a group of charges is the algebraic sum of the potentials for each of the various point charges.

势的叠加：任意点上由一组电荷引起的总电势是各个不同点电荷电势的代数和。

Potential difference is the difference in electric potential between two points.

势差：两点之间的电势差。

An **equipotential line** is a line along which the electrical potential is the same everywhere.

等势线：电势处处相同的线。

An **equipotential surface** is a surface over which the electrical potential is the same everywhere.

等势面：电势处处相同的面。

POTENTIAL DIFFERENCE

The potential difference between points A and B due to a point charge q depends only on the initial and final radial coordinates r_A and r_B. This is true for any static distribution of charge.

势差

点电荷 q 在 A 点和 B 点产生的势差只取决于初始和最终径向坐标 r_A 和 r_B。这一点对于任何静态电荷分布均成立。

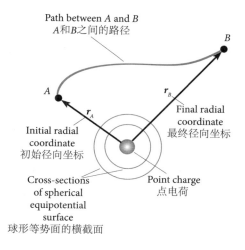

Path between A and B
A 和 B 之间的路径
Final radial coordinate
最终径向坐标
Initial radial coordinate
初始径向坐标
Point charge
点电荷
Cross-sections of spherical equipotential surface
球形等势面的横截面

Equipotential surface
等势面

Equipotential surfaces are perpendicular to the electric field lines
等势面垂直于电场线

Electric field line
电场线

EQUIPOTENTIAL SURFACE

Equipotential surfaces (blue lines) and electric field lines (grey lines) for a point charge.

等势面

点电荷的等势面（蓝线）和电场线（灰线）。

The electric field intensity is very high in the vicinity of a sharp point on a charged conductor. The field-ion microscope is a device that makes use of this intense field. Field-ion microscope image of the end of a sharp platinum needle. Each bright spot is a platinum atom.

带电导体的尖端附近电场强度非常高。场离子显微镜就是一种利用了这种强场的装置。右图为场离子显微镜下锋利的铂针尖端图像。每个亮点都是一个铂原子。

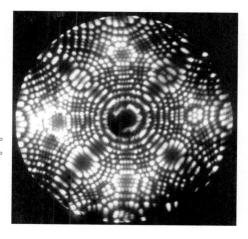

CAPACITANCE
电容

TERMS	术语
conductor	导体
electrically isolated conductor	绝缘导体
oppositely charged conductors	带相反电荷的导体
geometrical arrangement of conductor	导体的几何排布
charge on a conductor	导体上的电荷
stored charge	储存的电荷
potential difference between conductors	导体间电势差
material separating the charged conductors	分隔带电导体的材料
insulator	绝缘体
dielectric	电介质
capacitor	电容器

LANGUAGE OF PHYSICS   物理学用语

The **capacitance of a capacitor** is the ratio of the magnitude of the charge on either conductor to the magnitude of the potential difference between them.

电容器电容：两导体中任一导体上电荷大小与两导体间电势差大小之比。

The **capacitance of a device** is a measure of its ability to store charge and electrical potential energy.

装置电容：衡量该装置储存电荷和电势能的能力。

An **insulator** or **dielectric** is a material that does not permit the free flow of electric charge through it. Placing a dielectric between the plates of a capacitor increases the capacitance of that capacitor.

绝缘体或电介质：一种不允许电荷自由流动的材料。在电容器的两极板之间放置电介质可以增加电容器的电容。

CAPACITOR

A capacitor consists of two conductors electrically isolated from each other and their surroundings. Once the conductors are charged, the two conductors carry equal and opposite charges.

电容器

电容器由两个彼此绝缘且与周围环境也绝缘的导体构成。导体一旦被充电,这两个导体就带等量的异种电荷。

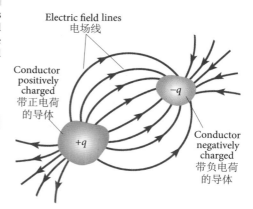

Electric field lines
电场线

Conductor positively charged
带正电荷的导体

$+q$

$-q$

Conductor negatively charged
带负电荷的导体

A flexible supercapacitor made in the National University "Lviv Polytechnic". A supercapacitor is a high-capacity electrochemical capacitor with capacitance values much higher than other capacitors (but lower voltage limits) that bridges the gap between electrolytic capacitors and rechargeable batteries. They typically store 10 to 100 times more energy per unit volume or mass than electrolytic capacitors, can accept and deliver charge much faster than batteries, and tolerate many more charge and discharge cycles than rechargeable batteries.

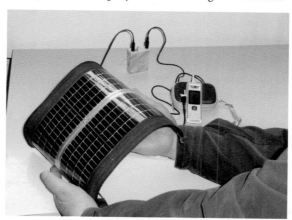

乌克兰利沃夫国立理工大学制造的柔性超级电容器。超级电容器是一种具有高电容的电化学电容器,其电容值远高于其他电容器(但临界电压较低),弥补了电解质电容器和可充电电池之间的差距。单位体积或质量的超级电容器通常比电解质电容器多存储10~100倍的能量,它们充放电比电池更快,并且其可承载的充放电循环次数也比可充电电池多得多。

CAPACITOR
电容器

TERMS 术语

capacitor	电容器
parallel-plate capacitor	平行板电容器
cylindrical capacitor	圆柱形电容器
spherical capacitor	球形电容器
capacitor with dielectric	填充介质的电容器
high-voltage capacitor	高压电容器
variable capacitor	可变电容器
electrolytic capacitor	电解质电容器
capacitance of capacitor	电容器电容
area of plate	极板面积
plate separation	极板间距
edge effect	边缘效应
negligible of edge effect	边缘效应可忽略不计
cylindrical shell	圆柱壳
spherical shell	球形壳
conducting shell	导电壳
coaxial	同轴
concentric	同心
dielectric material	介电材料
dielectric constant	介电常数
dielectric strength	介电强度
short circuit	短路

LANGUAGE OF PHYSICS 物理学用语

A **capacitor** is a device that accumulates and stores electrical energy.

电容器：一种积累和储存电能的装置。

A **capacitor** consists of two conductors of **any size or shape** carrying equal and opposite charges separated by an insulator (insulating medium).

电容器：由两个**任意大小或形状**的导体组成，携带等量异种电荷，由绝缘体（绝缘介质）隔开。

The **dielectric constant** is defined as the ratio of the capacitance with a dielectric between the plates to the capacitance with air or vacuum between them.

介电常数：两极板之间填充介质时的电容与两极板之间填充空气或真空时的电容之比。

Dielectric strength is the value of the potential difference per unit plate separation when the dielectric breaks down.

介电强度：介电击穿时每单位板间距的电势差值。

电磁学 **171**

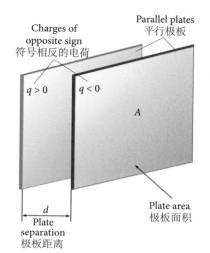

PARALLEL-PLATE CAPACITOR

A parallel-plate capacitor consists of two parallel plates each of area A, separated by a distance d. When the capacitor is charged, the plates carry equal charges of opposite sign.

平行板电容器

平行板电容器由两个面积为A、间距为d的平行极板构成。当电容器充电时，极板带等量异种电荷。

CYLINDRICAL CAPACITOR

A cylindrical capacitor consists of a cylindrical conductor of radius a and length l surrounded by a coaxial cylindrical shell of radius b. The field is confined to the region between cylinders.

圆柱形电容器

圆柱形电容器由半径为a、长度为l的圆柱形导体和将其包裹其中的半径为b的同轴圆柱壳组成。电场被局限于圆柱体之间的区域。

SPHERICAL CAPACITOR

A spherical capacitor consists of a spherical conductor of radius b and charge $-q$ that is concentric with a smaller conducting sphere of radius a and charge $+q$.

球形电容器

球形电容器由半径为b、电荷为$-q$的球形导体与半径为a、电荷为$+q$的较小同心导电球体组成。

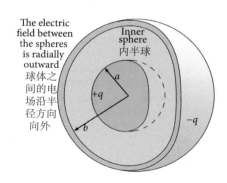

COMBINATION of CAPACITORS
电容器的连接

TERMS 术语	
combination	连接
parallel combination	并联
series combination	串联
equivalent capacitance	等效电容
equivalent capacitor	等效电容器
individual capacitance	单个电容
potential difference (voltage) across capacitor	电容器之间的电势差（电压）
terminal of battery	电池接线柱
positive terminal of battery	电池正极
negative terminal of battery	电池负极
total charge	总电荷
charge transfer process	电荷转移过程
discharge	放电
uncharged capacitor	未充电的电容器
visible spark	可见火花
conducting wire	导线

LANGUAGE OF PHYSICS 物理学用语

Capacitors in series

When capacitors are connected **in series** the reciprocal of the equivalent **capacitance** is equal to the sum of the reciprocals of each capacitor.
Each capacitor has the same magnitude of **charge** on its plates.
The total **potential** across capacitors is equal to the sum of the potential drops across each capacitor.

串联电容器

当电容器**串联**时，等效**电容**的倒数等于每个电容器的倒数之和。
每个电容器的极板上都有同样大小的**电荷**。
通过电容器的总**电势**等于通过每个电容器的电势降之和。

Capacitors in parallel

When capacitors are connected **in parallel** the equivalent **capacitance** is equal to the sum of the individual capacitances.
The **potential drop** across each capacitor is the same.
The total **charge** is equal to the sum of charges deposited on the plates of each capacitor.

并联电容器

当电容器**并联**时，等效**电容**等于单个电容的加和。
通过每个电容器的**电势降**是相同的。
总**电荷**等于在每个电容器的极板上积累的电荷之和。

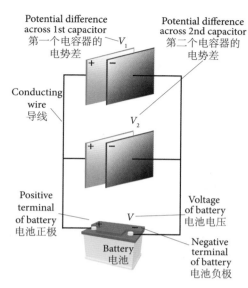

Potential difference across 1st capacitor 第一个电容器的电势差

Potential difference across 2nd capacitor 第二个电容器的电势差

Conducting wire 导线

Positive terminal of battery 电池正极

Voltage of battery 电池电压

Battery 电池

Negative terminal of battery 电池负极

PARALLEL CAPACITORS

A parallel combination of two capacitors. The equivalent capacitance of a parallel combination of capacitors is larger than any of the individual capacitances.

并联电容器

两个电容器的并联。并联电容器的等效电容比其中任何单个电容器的电容都要大。

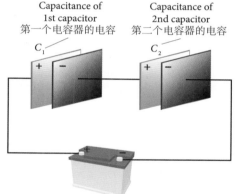

Capacitance of 1st capacitor 第一个电容器的电容

Capacitance of 2nd capacitor 第二个电容器的电容

CAPACITORS IN SERIES

A series combination of two capacitors. The equivalent capacitance of a series combination is always less than any individual capacitance in the combination.

串联电容器

两个电容器的串联。串联电容器的等效电容总是小于组合中的单个电容。

When the key on a computer keyboard is depressed, the spacing between the plates of a capacitor beneath the key changes, causing a change in capacitance. An electrical signal derived from this capacitance change is used to register the keystroke.

当电脑键盘上的键被按下时，键下电容器极板之间的距离就会改变，从而引起电容的变化。由电容变化产生的电信号则用于表达该按键。

ELECTRIC DIPOLE and POLARIZATION
电偶极子和电极化

TERMS 术语

electric dipole	电偶极子
partially aligned electric dipoles	部分平行的电偶极子
randomly oriented electric dipoles	随机取向的电偶极子
electric dipole moment	电偶极矩
induced moment	感应矩
equal charges	等量电荷
opposite charges	符号相反的电荷
charges separated by a distance	相隔一段距离的电荷
internal electric field	内电场
external electric field	外电场
electric field without dielectric	无介质电场
electric field in the presence of a dielectric	存在介质时的电场
torque	力矩
polar molecule	极性分子
nonpolar molecule	非极性分子
degree of alignment	排列一致的程度
polarization	极化
permanent polarization	永久极化
induced polarization	感应极化
surface charge	面电荷

LANGUAGE OF PHYSICS 物理学用语

An **electric dipole** consists of two equal and opposite charges separated by a distance.

电偶极子：由两个大小相等、符号相反、间隔一定距离的电荷组成。

The **electric dipole moment** is the product of the charge times the distance separating the charges.

电偶极矩：电荷大小与电荷之间距离的乘积。

ELECTRIC DIPOLE

The electric dipole moment p is directed from negative charge $-q$ to positive charge $+q$.

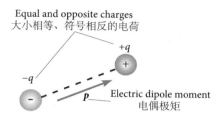

ELECTRIC DIPOLE IN A UNIFORM ELECTRIC FIELD

The dipole moment p is at an angle θ with the field, and the dipole experiences a torque.

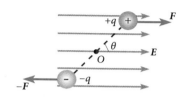

电偶极子
电偶极矩p从负电荷$-q$指向正电荷$+q$。

均匀电场中的电偶极子
电偶极矩p在与电场夹角为θ时，电偶极子受到一个力矩的作用。

MOLECULES WITH A PERMANENT DIPOLE MOMENT

Molecules with a permanent dipole moment are randomly oriented in the absence of an external electric field.

When an external field is applied, the dipoles are partially aligned with the field and the dielectric is polarized.

This polarization causes an induced negative surface charge on one side and an equal positive surface charge on the opposite side. This results in a reduction in the net electric field within the dielectric.

具有永久电偶极矩的分子

在没有外电场的情况下，具有永久电偶极矩的分子随机取向。

在外电场的作用下，电偶极子与电场部分平行，电介质被极化。

这种极化导致在一边产生负的表面感应电荷，在另一边产生等量正的表面感应电荷。因此，电介质内的净电场减小。

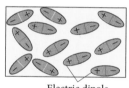

ELECTRIC CURRENT
电流

TERMS 术语

current	电流
average current	平均电流
instantaneous current	瞬时电流
direction of current	电流方向
steady current	稳恒电流
current in conductor	导体中的电流
current density	电流密度
flow of charges	电荷流
flow of electrons	电子流
flow of charged particles	带电粒子流
charge carrier	电荷载体
drift speed	漂移速度
electron gas	电子气
conduction electron	导电电子
surface of cross-sectional area	横截面
surface perpendicular to direction of current	垂直于电流方向的面

LANGUAGE OF PHYSICS 物理学用语

The **current** is the rate at which charge flows through a surface.

电流：电荷流过横截面的速率。

Electron current is the actual current in a circuit, it is a flow of electrons from a region of low potential to a region of high potential.

电子电流：电路中实际的电流，是从低电势区域到高电势区域的电子流。

The **direction of the current** is opposite to the direction of flow of electrons.

电流方向与电子流动的方向相反。

The **current density** in a conductor is defined to be the current per unit area.

导体中的**电流密度**：单位面积上的电流。

CURRENT DIRECTION | 电流方向

Charges in motion through an area A. The direction of the current is the direction in which positive charge would flow if free to do so.

通过横截面面积A的电荷。电流的方向是正电荷自由流动的方向。

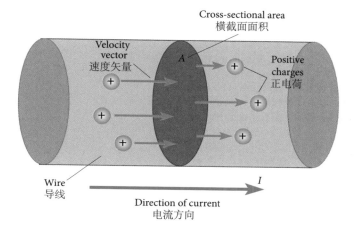

CHARGE MOTION | 电荷运动

The zigzag motion of a charge carrier in a conductor. The changes in direction are due to collisions with atoms in the conductor. The net motion of electrons is opposite to the direction of the electric field.

导体中电荷载体的"之"字运动。方向的变化是由于与导体中的原子发生碰撞。电子的净运动方向与电场的方向相反。

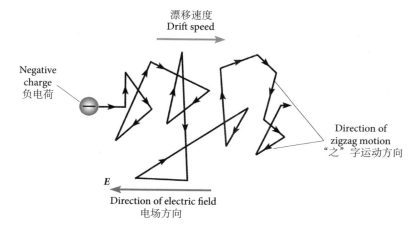

RESISTANCE and OHM'S LAW
电阻和欧姆定律

TERMS 术语	
resistance	电阻
conductivity	电导率
resistivity	电阻率
resistor	电阻器
circuit	电路
ohmic material	欧姆材料
nonohmic material	非欧姆材料
current-voltage relationship	电流-电压关系
linear current-voltage relationship	线性电流-电压关系
nonlinear current-voltage relationship	非线性电流-电压关系
current density	电流密度
conductor	导体
potential difference	电势差
wire	导线

LANGUAGE OF PHYSICS 物理学用语

The **resistance** of a conductor is directly proportional to its resistivity and its length, and inversely proportional to its cross-sectional area A.

导体**电阻**与导体电阻率和导体长度成正比，与导体横截面积A成反比。

A **resistor** is a device having electrical resistance and is used in an electric circuit for purposes of protection, operation or to control a current.

电阻器：一种具有电阻的装置，在电路中用于保护、操作或控制电流。

Ohm's law
For many materials (including most metals), the ratio of the current density to the electric field is a constant that is independent of the electric field producing the current.

欧姆定律
许多材料（包括大多数金属）的电流密度与电场的比值是一个常数，与产生电流的电场无关。

Joule heat is the energy that is lost as a charge passes through a resistor.

焦耳热：电荷通过电阻器时损失的能量。

OHMIC AND NONOHMIC MATERIALS

The current-voltage curve for an ohmic (a) and nonohmic (b) material.

欧姆材料和非欧姆材料

欧姆（a）和非欧姆（b）材料的电流-电压曲线。

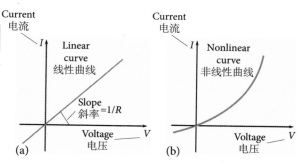

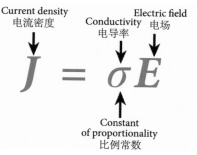

OHM'S LAW

A current density J and an electric field E are established in a conductor when a potential difference is maintained across the conductor.

欧姆定律

导体两端保持电势差时，导体中便有了电流密度 J 和电场 E。

The colored bands on a resistor represent a code for determining the value of its resistance. The 1st two colors give the 1st two digits in the resistance value. The 3rd color gives the power of 10 for the multiplier of the resistance value. The last color is the tolerance.

Resistors colour code
变阻器的颜色代码

Colour	Number 数字	Multiplier 乘数	Tolerance 公差	颜色
Black	0	1		黑色
Brown	1	10^1		棕色
Red	2	10^2		红色
Orange	3	10^3		橙色
Yellow	4	10^4		黄色
Green	5	10^5		绿色
Blue	6	10^6		蓝色
Violet	7	10^7		紫罗兰色
Gray	8	10^8		灰色
White	9	10^9		白色
Gold		10^{-1}	5%	金色
Silver		10^{-2}	10%	银色
Colourless			20%	无色

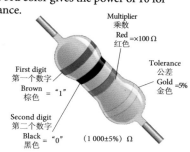

电阻器上的彩色带代表一个确定其电阻值的代码。前两种颜色表示电阻值的前两位数字。第三种颜色表示电阻值与10的乘数。最后一种颜色表示公差。

RESISTANCE and TEMPERATURE
电阻和温度

TERMS	术语
resistivity versus temperature	电阻率随温度的变化
temperature coefficient of resistivity	电阻率的温度系数
metal	金属
semiconductor	半导体
pure superconductor	纯超导体
high-temperature superconductor	高温超导体
critical temperature	临界温度
superconductivity	超导电性
resistivity drops to zero	电阻率降为零
resistivity varies linearly	电阻率呈线性变化
resistivity increases	电阻率增加
resistivity decreases	电阻率降低
resistivity approaches a finite value	电阻率趋于有限值
superconducting magnet	超导磁体
liquid nitrogen	液氮
BCS theory	BCS 理论

LANGUAGE OF PHYSICS
物理学用语

A **metal** is a material that has good electrical and thermal conductivity.

金属：一种具有良好导电性和导热性的材料。

A **semiconductor** is a material whose ability to conduct electricity falls between that of conductors and insulators. The conductivity of semiconductors can be controlled by adding specific amounts of impurities in a process called doping.

半导体：一种导电能力介于导体和绝缘体之间的材料。半导体的导电性可以通过在掺杂过程中加入特定量的杂质来控制。

A **superconductor** is a metal, alloy or compound that loses its electrical resistance at temperatures below a certain transition temperature referred to as the critical temperature T_c. High-temperature superconductors occur near 130 K, while low-temperature superconductors have T_c in the range of 4 to 18 K.

超导体：在低于某一转变温度（即临界温度T_c）时失去电阻的某种金属、合金或化合物。高温超导体出现在130 K附近，低温超导体的T_c则为4~18 K。

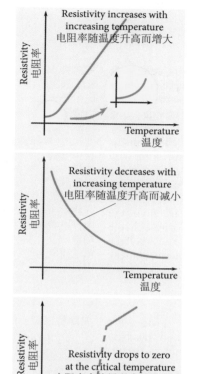

METAL
Resistivity versus temperature for a **metal**. As T approaches absolute zero, the resistivity approaches a finite value.

金属
金属电阻率随温度的变化。当 T 趋于绝对零度时，电阻率趋于有限值。

SEMICONDUCTOR
Resistivity versus temperature for a pure **semiconductor**.

半导体
纯**半导体**的电阻率随温度的变化。

SUPERCONDUCTOR
Resistivity versus temperature for a **superconductor**. The resistance drops to zero at the critical temperature.

超导体
超导体的电阻率随温度的变化。在临界温度以下电阻降到零。

Carbon atoms can form stable bonds in multiple ways, so pure carbon can be found naturally in many forms, such as diamond, graphite, soot, and nanoscopic structures such as tubes, fullerenes and cages. It can form a single layered hexagonal mesh known as graphene, which, although a good conductor, is not a superconductor. However, a double layer of graphene can mimic the resistanceless behaviour of a superconductor. In April 2018, a group at MIT in the U.S. showed that it is possible to generate a form of superconductivity in a system of two layers of graphene twisted by 1.1°.

碳原子能以多种方式形成稳定的键，所以纯碳能以多种形式在自然界中存在，如钻石、石墨、煤烟以及纳米结构（如纳米管、富勒烯和纳米笼）。它可以形成单层的六边形网格，即石墨烯。虽然石墨烯是良好的导体，但不是超导体。然而，双层石墨烯具有类似于超导体的无电阻特性。2018年4月，美国麻省理工学院（MIT）的一个团队证明，将双层石墨烯上下层相对扭转1.1°时可能产生某种超导电性。

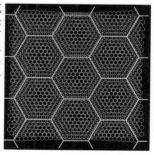

ELECTROMOTIVE FORCE
电动势

TERMS 术语	
emf (ee-em-ef)	电动势
source of energy	能量来源
source of emf	电动势来源
charge pump	电荷泵
battery	电池
battery connected to resistor	连接电阻器的电池
connecting wire	连接导线
resistance	电阻
internal resistance	内阻
load resistance	负载电阻
open-circuit voltage	开路电压
work done per unit charge	单位电荷做的功
terminal voltage	端电压
potential difference across battery	电池间电势差
circuit diagram	电路图

LANGUAGE OF PHYSICS   物理学用语

A **source of emf** is any device that produces an electric field and thus may cause charges to move around a circuit.

电动势来源：任何能产生电场且可以使电荷在电路中移动的装置。

A **battery** is a combination of electrolytic cells that supplies potential difference and in the process converts chemical energy to electrical energy.

电池：提供电位差并在此过程中将化学能转化为电能的电解池的组合。

The **emf of a battery** is equal to the voltage across its terminals when the current is zero.

电池电动势等于电流为零时电池两极的电压。

The **emf is** equivalent to the open circuit voltage of the battery.

电动势等于电池的开路电压。

EMF CIRCUIT DIAGRAM

Circuit diagram of a battery of emf ε and internal resistance r connected to an external resistor R. The battery within the dashed rectangle is represented by an emf in series with the internal resistance.

电动势电路图

具有电动势 ε 和内阻 r 的电池连接到外部电阻 R 的电路图。虚线矩形框内的电池用电动势和串联在一起的内阻来表示。

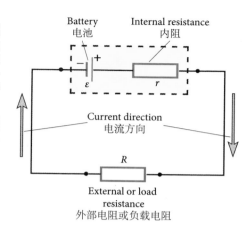

Battery 电池 Internal resistance 内阻
Current direction 电流方向
External or load resistance 外部电阻或负载电阻

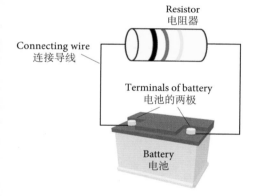

Resistor 电阻器
Connecting wire 连接导线
Terminals of battery 电池的两极
Battery 电池

CIRCUIT WITH BATTERY

A circuit consisting of a battery connected to a resistor.

带电池的电路

电路由连接电阻器的电池组成。

This is an innovative and practical AA rechargeable battery charger with a USB connector. You can plug it into a USB port on your computer for convenient charging, and the AA rechargeable battery takes 5-7 hours to fully charge. After fully charging the USB battery, with a capacity of 1 450 mAh, the USB battery offers energy for your camera, remote, flashlight and more.

这是一款新颖实用且带USB接口的5号充电电池充电器。您可以将其插入电脑的USB接口，给电脑方便地充电。充满一节五号电池需要5~7个小时。USB电池的容量为1 450毫安时，在完全充满电后，可以为您的相机、遥控器、手电筒和更多设备提供电源。

COMBINATION of RESISTORS
电阻器的连接

TERMS	术语
connection	连接
series connection	串联
parallel connection	并联
potential drop	电势降
resistance	电阻
individual resistance	单个电阻
equivalent resistance	等效电阻
potential difference across resistor	电阻器两端的电势差
junction	节点
lightbulb	灯泡

LANGUAGE OF PHYSICS 物理学用语

The **equivalent resistance** is the resistance of a single resistor which is equal to the combined resistance of the individual resistors in the circuit.

等效电阻：用单个电阻器的电阻等效地表示电路中各个电阻器连接后的总电阻。

A **series circuit** is a circuit in which each element of the circuit is connected to an adjacent element of the circuit so that the same current passes through each element.

串联电路：该电路中的每个元器件都与邻近元器件相连，且通过每个元器件的电流相同。

For **resistors in series**, the equivalent **resistance** is equal to the sum of the individual resistances. The **current** is the same through each resistor.

串联电阻器的等效电阻等于各个电阻之和。通过每个电阻的电流相同。

A **parallel circuit** is a circuit in which the circuit elements are connected in such a way that the potential difference across all the elements of the circuit is the same.

并联电路：该电路中元器件的连接方式使电路中的所有元器件之间的电势差相同。

For **resistors in parallel**, the reciprocal of the equivalent **resistance** is equal to the sum of the reciprocals of the individual resistances. The **potential difference** across the resistors is the same.

并联电阻器：其等效电阻的倒数等于各个电阻的倒数之和。每个电阻器两端的电势差相同。

SERIES CONNECTION

Series connection of two resistors, R_1 and R_2. The current is the same in each resistor. The same amount of charge flows through each circuit element.

串联

电阻R_1和R_2串联。通过每个电阻器的电流相同。流过每个电路元器件的电荷数量相同。

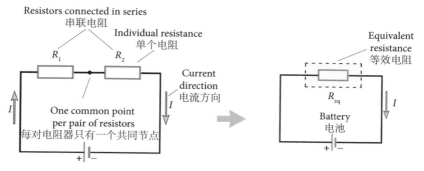

PARALLEL CONNECTION

Parallel connection of two resistors, R_1 and R_2. The potential difference across each resistor is the same.

并联

电阻R_1和R_2并联。每个电阻器两端的电势差相同。

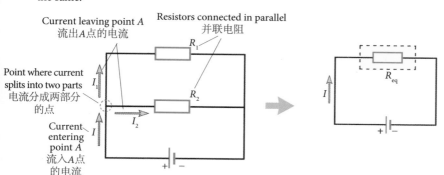

The lightbulbs for the light set on a tram are connected in parallel, so that if one fails, the others remain on.

有轨电车上的灯泡是并联的，所以如果其中一个坏了，其他的仍然亮着。

KIRCHHOFF'S RULES
基尔霍夫定律

TERMS	术语
circuit	电流
loop	回路
single-loop circuit	单回路电路
multiloop circuit	多回路电路
closed circuit	闭合电路
junction	节点
current entering junction	流入节点的电流
current leaving junction	流出节点的电流
potential drop	电势降
resistor traversed in the direction of the current	沿电流方向被穿过的电阻器
resistor traversed in the direction opposite to the current	沿电流反方向被穿过的电阻器
emf traversed in the direction of the emf	沿电动势方向被穿过的电动势
emf traversed in the direction opposite to the emf	沿电动势反方向被穿过的电动势
junction rule	节点定律
loop rule	回路定律
number of independent equations	独立方程数
matrix algebra	矩阵代数

LANGUAGE OF PHYSICS
物理学用语

A **junction** is any point in a circuit where a current can split.

节点：电路中可以将电流分开的任意一点。

Kirchhoff's rules
1. The sum of the currents entering any junction must equal the sum of the currents leaving that junction.

2. The algebraic sum of the changes in potential across all of the elements around any closed circuit loop must be zero.

基尔霍夫定律
1.流入某一节点的电流之和必须等于流出该节点的电流之和。

2.在任何闭合回路中，通过所有元器件的电势变化的代数和必须为零。

JUNCTION RULE

A diagram illustrating Kirchhoff's junction rule. All currents entering a junction must leave that junction.

节点定律

基尔霍夫节点定律示意图。所有流入一个节点的电流必须流出该节点。

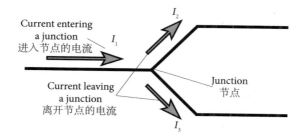

SINGLE-LOOP CIRCUIT

An example of single-loop circuit containing two resistors and two batteries connected in series. The polarities of the batteries are in opposition to each other.

单回路电流

由两个电阻和两个电池串联而成的单回路电路的例子。两个电池的电极反向连接。

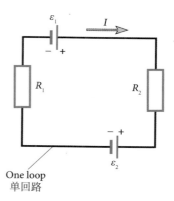

MULTILOOP CIRCUIT

An example of multiloop circuit containing three loops *abefa*, *bcdeb*, *acdfa* and two junctions *b*, *e*.

多回路电流

多回路电路的例子，包含三条回路 *abefa*、*bcdeb*、*acdfa* 和两个节点 *b* 和 *e*。

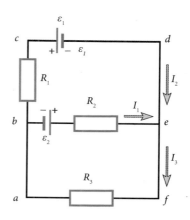

ELECTRICAL INSTRUMENT
电气仪表

TERMS	术语
ammeter	电流表
voltmeter	电压表
galvanometer	电流计
Wheatstone bridge	惠斯通电桥
potentiometer	电位器
digital multimeter	数字万用表
resistance	电阻
zero resistance	零电阻
infinite resistance	无穷大电阻
shunt	分流器
connected in parallel	并联
connected in series	串联
positive terminal	正极

LANGUAGE OF PHYSICS 物理学用语

An **ammeter** is a device that measures the amount of current in a circuit.

电流表：测量电路中电流大小的一种装置。

A **voltmeter** is a device that measures the potential difference between any two points in an electric circuit.

电压表：测量电路中任意两点之间电位差的装置。

A **galvanometer** is a device that indicates that there is a current in a circuit.

电流计：指示电路中有电流的一种装置。

A **Wheatstone bridge** is a circuit for measuring an unknown resistance.

惠斯通电桥：一种测量未知电阻的电路。

A **potentiometer** is a circuit used to measure an unknown emf by comparison with a known emf.

电位器：通过与已知电动势比较来测量未知电动势的电路。

AMMETER

The current in a circuit can be measured with an ammeter connected in series with the resistor and battery. An ideal ammeter has zero resistance.

电流表

电路中的电流可以用与电阻器和电池串联的电流表来测量。理想的安培表电阻为零。

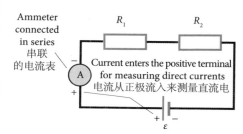

Ammeter connected in series
串联的电流表

Current enters the positive terminal for measuring direct currents
电流从正极流入来测量直流电

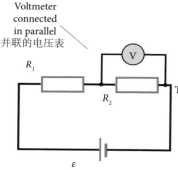

Voltmeter connected in parallel
并联的电压表

The positive terminal connected to the end of resistor at higher potential
正极连接在电阻器的高电势端

VOLTMETER

The potential difference across a resistor can be measured with a voltmeter connected in parallel with a resistor. An ideal voltmeter has infinite resistance.

电压表

通过电阻器的电势差可以用并联在电阻器上的电压表来测量。理想的电压表电阻无穷大。

WHEATSTONE BRIDGE

Circuit diagram for a Wheatstone bridge. This circuit is used to measure an unknown resistance R_x in terms of known resistances R_1, R_2 and R_3.

惠斯通电桥

惠斯通电桥电路图。该电路可以用已知电阻 R_1、R_2 和 R_3 来测量一个未知电阻 R_x。

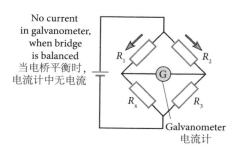

No current in galvanometer, when bridge is balanced
当电桥平衡时，电流计中无电流

Galvanometer
电流计

Voltages, currents, and resistances can be measured using digital multimeters.

电压、电流和电阻都可以用数字万用表测量。

MAGNETIC FIELD
磁场

TERMS	术语
Oersted's experiment	奥斯特实验
compass needle	指南针
magnet	磁体
bar magnet	条形磁铁
horseshoe magnet	马蹄磁铁
permanent magnet	永磁体
pole	极
north pole	北极
south pole	南极
iron filing	铁屑
moving charged particle	移动的带电粒子
current-carrying conductor	通电导体
magnetic field	磁场
magnitude or strength of magnetic field	磁场大小或磁场强度
direction of the magnetic field	磁场方向
steady magnetic field	稳恒磁场
uniform magnetic field	均匀磁场
magnetic field line	磁场线
magnetic induction	磁感应强度
magnetic field density	磁场密度

LANGUAGE OF PHYSICS   物理学用语

A **magnetic field** is the area around a magnet, moving electric charges or electric current, or changing electric field, in which there is a magnetic force.

磁场：存在于磁体、运动的电荷或电流，或者交变电场周围的区域。在磁场中存在磁力。

Magnetic induction is equal to the force per unit charge per unit velocity that acts on a charge that is moving perpendicular to the magnetic field.

磁感应强度：等于单位电荷单位速度作用在垂直于磁场运动的电荷上的力。

Fundamental principle of magnetism
Like magnetic poles repel each other, unlike magnetic poles attract each other.

磁学基本原则
同性磁极相互排斥，异性磁极相互吸引。

MAGNETIC FIELD LINES

Several magnetic field lines of a bar magnet. The direction of the magnetic field B at any location is in the direction in which the north pole of a compass needle points at that location.

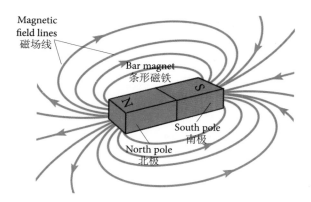

磁场线

条形磁铁的几条磁场线。在任何位置上，磁场 B 的方向与指南针的北极所指向的方向一致。

Magnetic field patterns between like poles of two bar magnets (a) and unlike poles of two bar magnets (b) as displayed with iron filings

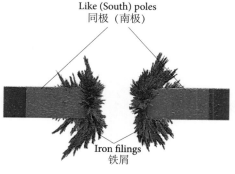

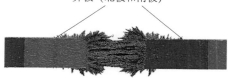

由铁屑展示出的两块条形磁铁同极（a）和异极（b）间的磁场样式。

MOTION of a CHARGED PARTICLE
带电粒子的运动

TERMS 术语

magnetic force	磁场力
uniform magnetic field	均匀磁场
nonuniform magnetic field	非均匀磁场
charged particle	带电粒子
moving charged particle	移动的带电粒子
rotating charged particle	旋转的带电粒子
circular path	圆周轨迹
helical path	螺旋轨迹
period of rotation	旋转周期
angular frequency	角频率
revolution	旋转
centripetal acceleration	向心加速度
radius of path	路径半径
cyclotron	回旋加速器
cyclotron frequency	回旋频率
dee	D形盒
magnetic bottle	磁瓶
velocity selector	速度选择器
mass spectrometer	质谱仪
Lorentz force	洛伦兹力
right-hand rule	右手定则

LANGUAGE OF PHYSICS 物理学用语

The **Lorentz force** is the total force acting on a moving charged particle and is composed of an electric force and a magnetic force. The magnetic field can only change the direction of the velocity of the moving charged particle but not its speed.

洛伦兹力：作用在运动的带电粒子上的总的力，包括一个电场力和一个磁场力。磁场只能改变带电粒子的运动速度的方向而不能改变其大小。

The **right-hand rule** (determines the direction of magnetic force)
Using your right-hand: point your index finger in the direction of the positive charge's velocity v. Point your middle finger in the direction of the magnetic field B. Your thumb now points in the direction of the magnetic force F.

右手定则（确定磁场力的方向）
用右手：食指指向正电荷运动速度v的方向，中指指向磁场B的方向，此时拇指所指的方向即为磁场力F的方向。

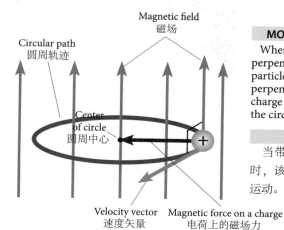

MOTION PERPENDICULAR TO A FIELD

When the velocity of a charged particle is perpendicular to a uniform magnetic field, the particle moves in a circular path whose plane is perpendicular to B. The magnetic force F on a charge is always directed toward the center of the circle.

垂直于磁场的运动

当带电粒子的运动速度垂直于均匀磁场时，该粒子将在垂直于磁场 B 的平面做圆周运动。电荷所受的磁场力 F 总是指向圆心。

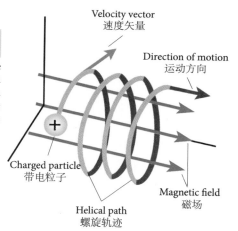

MOTION AT SOME ANGLE TO A FIELD

If a charged particle moves in a uniform magnetic field with its velocity at some arbitrary angle to B, its path is a helix.

与磁场成任意角度的运动

如果一个带电粒子在均匀磁场中的运动速度与 B 成一定角度，则其运动轨迹将呈螺旋状。

NONUNIFORM MAGNETIC FIELD

A charged particle moving in a nonuniform magnetic field spirals about the field and oscillates between the end points. This is called a magnetic bottle.

非均匀磁场

在非均匀磁场中运动的带电粒子绕磁场旋转并在端点之间振荡。该磁场被称为磁瓶。

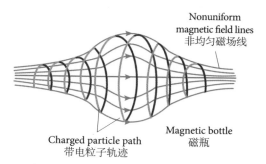

CURRENT-CARRYING CONDUCTOR
通电导体

TERMS 术语

current-carrying conductor	通电导体
current-carrying wire	通电导线
current-carrying loop	通电线圈
closed current-carrying conductor	闭合通电导体
wire segment	一段导线
straight wire segment	一段直导线
arbitrarily shaped wire segment	一段任意形状的导线
torque on current loop	通电线圈上的力矩
rectangular loop	矩形线圈
plane of loop	线圈平面
force couple	力偶
twisted loop	扭转的线圈
torque about any point	任意点的力矩
moment arm	力矩臂
magnetic moment directed out of (into) page	指向纸面外（里）的磁矩
cross	叉乘
dot	点乘
tail of arrow	箭尾
tip of arrow	箭头
loop rotation clockwise	线圈顺时针旋转
loop rotation counterclockwise	线圈逆时针旋转

LANGUAGE OF PHYSICS 物理学用语

A wire of length l carrying a current I in an external magnetic field B experiences a **force** given by the product of the magnitudes of B, I, l and $\sin\theta$, where θ is the angle between the current direction and B.

一根长度为l的导线，通有电流I，在磁场B的作用下受到一个**力**的作用。该力的大小等于B、I、l和$\sin\theta$的乘积，其中θ为电流方向与B之间的夹角。

The right-hand grip rule

To determine the direction of the magnetic field around a wire carying a current, grasp the wire with the right hand, with the thumb in the direction of the current, then the fingers will curl around the wire in the direction of the magnetic field.

右手定则

为了确定通电导线周围磁场的方向，用右手握住导线，大拇指指向电流方向，则四指绕导线弯曲的方向即为磁场方向。

WIRE IN A MAGNETIC FIELD

(a) When there is no current in the wire, it remains vertical. (b) When the current is upward, the wire deflects to the left. (c) When the current is downward, the wire deflects to the right.

磁场中的导线

（a）当导线中没有电流时，导线呈垂直状态。（b）当电流向上时，导线向左弯曲。（c）当电流向下时，导线向右弯曲。

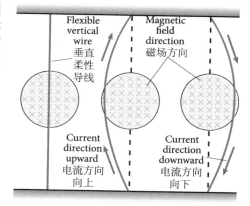

A blue cross indicates the direction of **B** is into the page. A blue dot indicates the direction of **B** is out of the page.

蓝色的叉表示磁场**B**的方向指向纸内。蓝色的点表示磁场**B**的方向指向纸外。

CURRENT LOOP IN MAGNETIC FIELD

A rectangular current loop in a uniform magnetic field. There are no forces acting on the sides parallel to **B**. The forces acting on the sides perpendicular to **B** create a torque that tends to twist the loop as shown. This fact has a number of important applications such as a DC motor, a galvanometer, an ammeter and a voltmeter.

磁场中的通电线圈

均匀磁场中的矩形通电线圈。线圈的两条边在平行于磁场**B**的方向不受力。而与磁场**B**垂直的两条边受到的力产生一个力矩，使线圈按如图所示方向旋转。许多重要应用由此而诞生，如直流电机、电流计、电流表和电压表。

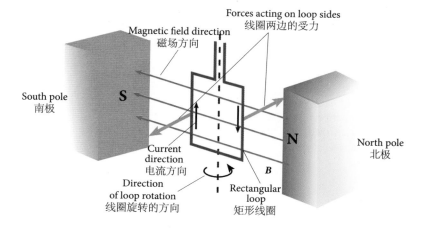

HALL EFFECT
霍尔效应

TERMS	术语
charge carrier	载流子
sign of the charge carrier	载流子符号
density of the charge carrier	载流子密度
drift speed	漂移速度
conductor	导体
semiconductor	半导体
flat strip	平的条带
upper (lower) edge of the flat strip	平条带的上（下）边缘
accumulation of charge	电荷积累
Hall voltage	霍尔电压
polarity	极性
Hall field	霍尔场
Hall coefficient	霍尔系数
quantum Hall effect	量子霍尔效应
Hall effect current sensor	霍尔效应电流传感器
magnetometer	磁力计

LANGUAGE OF PHYSICS
物理学用语

The production of a voltage difference (the Hall voltage) across a current carrying conductor in the presence of a magnetic field, perpendicular to both current and the magnetic field, is known as the **Hall effect**.

当有磁场存在时，导体中的电流会产生一个横向的电压差（霍尔电压），它既垂直于电流又垂直于磁场，该现象被称为**霍尔效应**。

The measured Hall voltage gives a value for the **drift speed.**

通过测量霍尔电压可以得到**漂移速度**的值。

The **charge density** is defined as the number of charge carriers per unit volume. It can be obtained by measuring the current in the conductor.

电荷密度：单位体积的载流子数目。该值能够通过测量导体中的电流得出。

A **Hall effect sensor** is a device that is used to measure the magnitude of a magnetic field. Its output voltage is directly proportional to the magnetic field strength through it. Often used as a magnetometer.

霍尔效应传感器：一种测量磁场强度大小的装置。该装置输出电压与磁场强度成正比。该装置通常被用作磁力计。

HALL VOLTAGE

When a current-carrying conductor is placed in a magnetic field, a voltage is generated in a direction perpendicular to both the current and the magnetic field. This arises from the deflection of charge carriers to one side of the conductor as a result of the magnetic force they experience. A voltmeter connected across the sample can be used to measure the potential difference generated across the conductor.

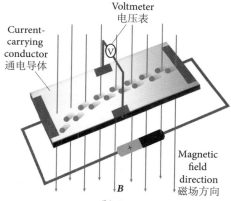

(a) Current-carrying conductor
通电导体

霍尔电压

当一个通电导体被放置于磁场中时,会产生一个既垂直于电流方向又垂直于磁场方向的电压。该现象产生的原因是载流子在磁场力的作用下向着导体一端偏转。 将电压表横向连接在导体上,就可以测量穿过导体的电势差。

(b) Current-carrying conductor in the magnetic field
磁场中的通电导体

Hall effect current sensor used to measure current. It may be used as a rotation speed sensor (for bicycle wheels, gear-teeth, automotive speedometers), fluid flow sensor, pressure sensor, electric airsoft gun, a trigger for electropneumatic paintball guns, and smart phones.

用于测量电流的霍尔效应电流传感器。它也可以用作(自行车轮、齿轮、汽车速度计中的)转速传感器、流量传感器、压强传感器、电动气枪、电动气动彩弹枪扳机以及智能手机。

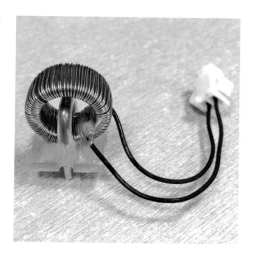

BIOT-SAVART LAW
毕奥-萨伐尔定律

TERMS	术语
steady current	稳恒电流
wire element	导线元
length of the wire element	导线元的长度
current element	电流元
current-carrying conductor	通电导体
conductor of finite size	有限大的导体
permeability of free space	真空磁导率
magnetic field at a point	某一点的磁场
total magnetic field	总磁场
long wire	长导线
straight wire	直导线
sum up	求和
integrate	积分
integrand	被积函数

LANGUAGE OF PHYSICS
物理学用语

The Biot-Savart law

The magnetic element dB at a point P due to a current element carrying a steady current I is proportional to the current, to the length ds of the element and to $\sin\theta$ (where θ is the angle between the vector ds and unit vector r directed from the element to that point), and inversely proportional to the square of the distance from the element to that point.

To find the **total magnetic field B** created at some point by a conductor of finite size, we must sum up contributions from all current elements making up the conductor.

毕奥-萨伐尔定律

一电流元承载稳恒电流 I，其在点 P 处产生的磁感应强度元 dB 与电流、基元长度 ds 以及 $\sin\theta$ 成正比（其中 θ 是矢量 ds 和从基元指向 P 点的单位矢量 r 之间的夹角），同时与基元到 P 点的距离的平方成反比。

为了求出一个有限大的导体在某点处产生的**总磁场 B**，我们需要将构成该导体的所有电流元产生的磁场累加起来。

MAGNETIC FIELD OF A CURRENT-CARRYING CONDUCTOR

The magnetic field d*B* at a point *P* due to a current element is given by the Biot-Savart law. The magnetic field is into the page at *P*.

通电导体的磁场

根据毕奥-萨伐尔定律，可以求出电流元在*P*点产生的磁场d*B*。该磁场在*P*点处指向纸内。

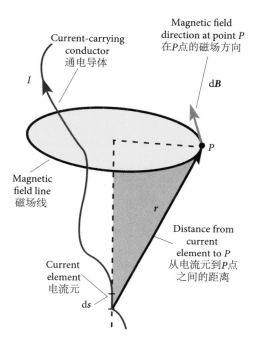

BIOT-SAVART LAW

The Biot-Savart law gives the magnetic field at a point only for a small element of the conductor. To find the total field at some point due to a current-carrying conductor, we must integrate this expression over the entire conductor.

毕奥–萨伐尔定律

毕奥-萨伐尔定律给出了导体中一个小的体积元在某一点处的磁场。要得到通电导体在某点处的总磁场，我们需要求出整个导体的积分。

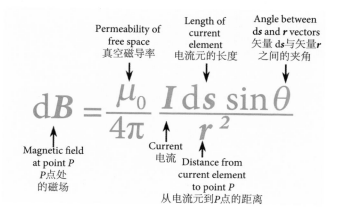

$$dB = \frac{\mu_0}{4\pi} \frac{I\,ds\,\sin\theta}{r^2}$$

AMPÈRE'S LAW
安培定律

TERMS	术语
right-hand rule	右手定则
circular path	圆形路径
closed circular path	闭合的圆形路径
circular path centered on the wire	以导线为中心的圆形路径
line integral	线积分
circumference of circle	圆周长
total current	总电流
steady current	稳恒电流
conduction current	传导电流
surface bounded by closed path	由闭合路径决定的平面
generalized form of Ampère's law	广义安培定律
changing electric field	变化电场

LANGUAGE OF PHYSICS 物理学用语

The right-hand grip rule
If a wire is grasped in the right hand with the thumb in the direction of the current, the fingers curl in the direction of B.

右手定则
用右手握住导线，大拇指指向电流方向，那么剩余四指弯曲方向即为磁场B的方向。

Ampère's law
The line integral of $B \cdot ds$ around any closed path equals $\mu_0 I$, where I is the total steady current passing through any surface bounded by the closed path.

安培定律
$B \cdot ds$绕任意闭合回路的线积分都等于$\mu_0 I$，其中I是穿过由闭合路径决定的平面的总稳恒电流。

Ampère's law is **valid only** for steady currents and is useful only for calculating the magnetic field of current configurations having a high degree of symmetry (long straight wire, solenoid, toroid).

安培定律只对稳恒电流**有效**，并且只适用于电流结构高度对称（如长直导线、螺线管或螺线环）的磁场。

The **generalized form of Ampère's law** describes the fact that magnetic fields are produced both by conduction currents and by changing electric fields.

广义安培定律表明传导电流和变化的电场都可以产生磁场。

MAGNETIC FIELD OF A CURRENT-CARRYING CONDUCTOR

A current-carrying conductor produces a magnetic field. To determine the direction of the magnetic field surrounding a long, straight wire carrying a current, one can use the right-hand rule.

通电导体的磁场

通电导体可以产生磁场。对于一根通电的长直导线，我们可以使用右手定则确定围绕该导线的磁场方向。

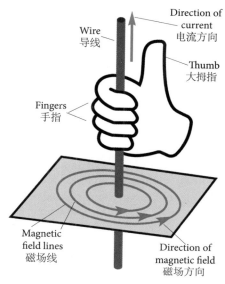

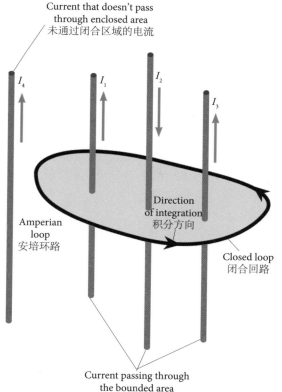

TOTAL CURRENT

The total current is the net current encircled by the loop. It can be found as the algebraic sum of all currents.

总电流

总电流是穿过闭合回路的净电流，其大小等于所有电流的代数和。

MAGNETIC FIELD of a SOLENOID
螺线管的磁场

TERMS	术语
solenoid	螺线管
loosely (tightly) wound solenoid	稀疏（紧密）螺线管
solenoid of finite size	有限尺寸螺线管
long wire	长导线
helix form	螺旋形式
turn of wire	线匝
closely spaced turn of wire	紧密缠绕的线匝
number of turns of wire	导线匝数
total number of turns of wire	导线总匝数
number of turns of wire per unit length	单位长度的导线总匝数
uniform magnetic field	均匀磁场
net magnetic field	净磁场
weak magnetic field	弱磁场
strong magnetic field	强磁场
circular loop	圆环
coil	线圈
magnetic field line	磁场线
magnetic field lines diverge from one end	磁场线在一端发散
magnetic field lines converge at opposite end	磁场线在另一端聚拢
toroid	螺线环
coil of ring-shaped form	环形线圈

LANGUAGE OF PHYSICS　　　　　　物理学用语

A **solenoid** is a coil whose length is substantially greater than its diameter, often wrapped around a metallic core, which produces a uniform magnetic field in a volume of space when an electric current is passed through it.

螺线管：一种长度远大于直径的线圈，通常缠绕在金属芯上，当通有电流时，能在一定空间产生一个均匀的磁场。

A **toroid** is a coil of insulated wire wound on a donut-shaped (ring-shaped) form made of powdered iron.

螺线环：一种绕在由铁粉制成的圆环上的绝缘线圈。

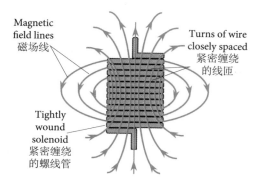

TIGHTLY WOUND SOLENOID

Magnetic field lines for a tightly wound solenoid of finite length carrying a steady current. The field in the space enclosed by the solenoid is nearly uniform and strong.

紧密缠绕的螺线管

通有稳恒电流、长度有限且紧密缠绕的螺线管产生的磁场线。螺线管内部的磁场几乎是均匀的，而且很强。

LOOSELY WOUND SOLENOID

The magnetic field lines for a loosely wound solenoid. The field lines between the turns tend to cancel each other. The field lines in the space surrounded by the coil are nearly parallel and uniformly distributed indicating that the field in this space is uniform.

稀疏缠绕的螺线管

稀疏缠绕的螺线管产生的磁场线。线匝之间的磁场线有相互抵消的趋势。线圈内部的磁场线几乎彼此平行且分布均匀，说明此空间中的磁场是均匀的。

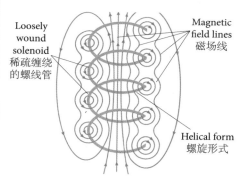

Magnetic resonance imaging (MRI) is a medical diagnostic technique. MRI scanners use a strong magnetic field produced by a superconducting solenoid.

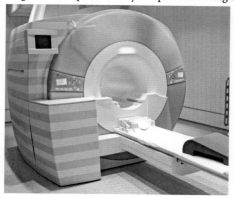
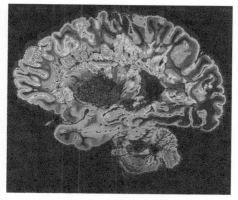

核磁共振成像（MRI）是一种医疗诊断技术，MRI扫描仪利用的是由超导螺线管产生的强磁场。

MAGNETIC FLUX
磁通量

TERMS	术语
magnetic flux	磁通量
total magnetic flux	总磁通量
magnetic flux through an area element	通过面积元的磁通量
magnetic flux through a closed surface	通过闭合表面的磁通量
surface of area A	面积为A的平面
net magnetic flux	净磁通量
density of the magnetic flux	磁通量密度
number of magnetic field lines entering surface	流入表面的磁场线数量
number of magnetic field lines leaving surface	流出表面的磁场线数量
monopole	磁单极子
isolated magnetic pole	孤立磁极
Gauss' law for magnetism	磁场的高斯定律
surface integral	面积分

LANGUAGE OF PHYSICS　　物理学用语

The **flux** is a quantitative measure of the number of lines of a vector field that passes perpendicularly through a surface.

通量：矢量场垂直穿过某一表面的场线数量的量化标准。

The **magnetic flux** is a quantitative measure of the number of lines of B that pass normally through a surface area A.

磁通量：磁场B垂直穿过一个面积为A的表面的场线数量的量化标准。

The **magnetic flux** through a surface is the surface integral of the normal component of the magnetic field B passing through that surface.

通过某一表面的**磁通量**等于穿过该表面的磁场B沿表面法向方向上的分量在该表面上的面积分。

Gauss's law for magnetism
The net magnetic flux through any closed surface is always zero.

磁场的高斯定律
穿过任意封闭表面的净磁通量总是零。

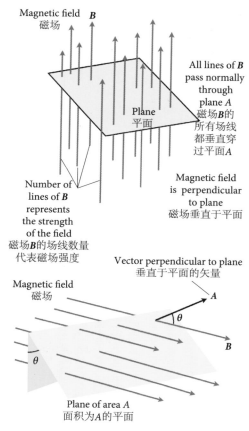

PLANE PERPENDICULAR TO THE FIELD

The flux through the plane is a maximum when the magnetic field is perpendicular to the plane.

垂直于磁场的平面

当磁场垂直于平面时,穿过平面的磁通量最大。

PLANE AT SOME ANGLE TO THE FIELD

A plane of area A and a uniform field B make an angle θ with the vector A. The magnetic flux through the plane is $BA\cos\theta$.

与磁场成一定夹角的平面

面积为A的平面与均匀磁场B之间的夹角等于B与矢量A之间的夹角θ。通过平面的磁通量大小为$BA\cos\theta$。

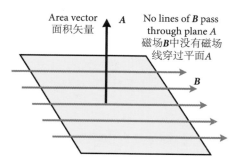

PLANE PARALLEL TO THE FIELD

The flux through the plane is zero when the magnetic field is parallel to the surface of the plane.

与磁场平行的平面

当磁场与平面表面平行时,穿过平面的磁通量为零。

MAGNETISM in MATTER
物质的磁性

TERMS	术语
classical model of an atom	经典原子模型
orbiting electron	在轨运行的电子
orbital motion	轨道运动
circular orbit	圆形轨道
area of the circular orbit	圆形轨道面积
radius of the circular orbit	圆形轨道半径
current loop	电流环
magnetic moment	磁矩
orbital magnetic moment	轨道磁矩
spin magnetic moment	自旋磁矩
angular momentum	角动量
orbital angular momentum	轨道角动量
spin angular momentum	自旋角动量
spin	自旋
Bohr magneton	玻尔磁子
magnetization	磁化强度
magnetization of a substance	物质的磁化强度
magnetization vector	磁化强度矢量
intrinsic magnetization	内禀磁化强度
magnetic field strength	磁场强度
magnetic susceptibility	磁化率
magnetic permeability	磁导率
diamagnetic	抗磁
paramagnetic	顺磁
ferromagnetic	铁磁

LANGUAGE OF PHYSICS   物理学用语

The **magnetic moments** in a magnetized substance may be described as arising from internal currents on the atomic level.

磁化物质中的**磁矩**可以认为是原子层面上的内部电流产生的。

The magnitude of the **magnetization vector** is equal to the magnetic moment per unit volume of the substance.

磁化强度矢量的大小等于单位体积物质的磁矩大小。

Diamagnetic materials are those whose atoms have no permanent magnetic moments.

抗磁材料：此类物质的原子没有永久磁矩。

Paramagnetic and **ferromagnetic materials** are those that have atoms with a permanent magnetic moment.

顺磁和铁磁材料：此类材料的原子具有永久磁矩。

MAGNETIC MOMENT OF AN ATOM

An electron moves in a circular orbit around the much more massive nucleus and has an orbital angular momentum. An orbiting electron constitutes a tiny current loop. A magnetic moment is associated with this current loop. The orbital angular momentum and magnetic moment are proportional to each other and in opposite directions.

原子磁矩

电子围绕着比其质量大很多的原子核做圆周运动，具有轨道角动量。一个绕轨运行的电子形成了一个微小的电流环，而磁矩的产生则与该电流环有关。轨道角动量和磁矩大小互相成正比，方向相反。

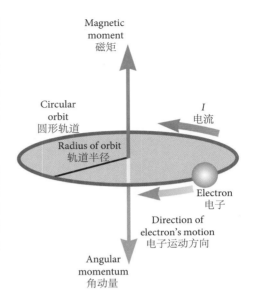

SPIN

An electron has an intrinsic property called spin, that also contributes to the magnetic moment. The electron behaves as if it were spinning about an axis through its center. This electron spin generates a magnetic field, the direction of which depends on the direction of the spin.

自旋

自旋是电子的一个内禀属性，也会产生磁矩。电子就好像在绕着自身的中心轴旋转。电子自旋会产生磁场，该磁场的方向取决于自旋的方向。

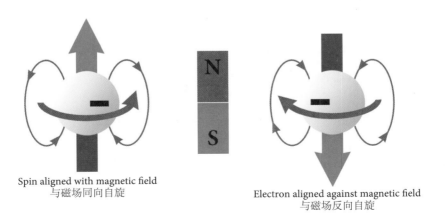

Spin aligned with magnetic field
与磁场同向自旋

Electron aligned against magnetic field
与磁场反向自旋

FERROMAGNETISM
铁磁性

TERMS 术语

substance	物质
crystalline substance	晶体物质
magnetized substance	磁化物质
unmagnetized substance	未磁化物质
demagnetized substance	退磁物质
ferromagnetic substance	铁磁性物质
magnetic dipole moment	磁偶极矩
permanent magnetic dipole moment	永久磁偶极矩
aligned magnetic dipole moment	平行的磁偶极矩
microscopic region	微观区域
domain	磁畴
domain randomly oriented	随机排列的磁畴
domain aligned with field	与磁场平行的磁畴
domain reoriented	重新排列的磁畴
domain wall	磁畴壁
saturation	饱和
magnetization curve	磁化强度曲线
B versus H curve	B-H曲线
magnetic hysteresis	磁滞
hysteresis loop	磁滞回线
remanent magnetization	剩余磁化强度
ferromagnet	铁磁体
soft ferromagnet	软铁磁体
hard ferromagnet	硬铁磁体
Curie temperature	居里温度

LANGUAGE OF PHYSICS 物理学用语

A **ferromagnet** is a substance, that below a certain temperature, the Curie point, can possess magnetization in the absence of an external magnetic field.

铁磁体：在一定温度（居里温度）以下可以在没有外加磁场的情况下磁化的一种物质。

Ferromagnetism is the property of certain materials that causes them to have relative permeabilities that exceed unity. This permeability permits the materials to exhibit hysteresis.

铁磁性：特定物质具有的性质，该性质使物质的相对磁导率大于1，从而使材料展现出磁滞现象。

DOMAIN STRUCTURE

(a) Random orientation of atomic magnetic dipoles in an unmagnetized substance. (b) When an external field is applied, the atomic magnetic dipoles tend to align with the field, giving the sample a net magnetization.

磁畴结构

(a) 未磁化物质中原子磁偶极子的随机取向。(b) 在外磁场下，原子磁偶极子倾向于与外场平行，使得样品具有净磁化强度。

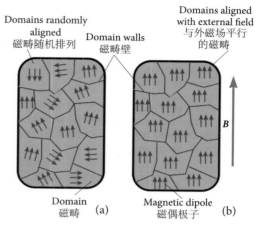

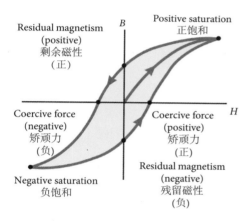

HYSTERESIS

A magnetic hysteresis loop is the graphical representation of the magnetic memory of ferromagnetic materials. A ferromagnetic substance remains magnetized after the external field is removed.

磁滞

磁滞回线以图示方式表明铁磁材料具有磁记忆。撤掉外场后，铁磁物质仍然是被磁化的。

HARD AND SOFT FERROMAGNETS

(a) Hysteresis curve for a hard ferromagnetic material corresponding to a large remanent magnetization. Such materials cannot be easily demagnetized by an external field and are used in permanent magnets. (b) Hysteresis curve for a soft ferromagnetic material corresponding to a small remanent magnetization. Such materials are easily magnetized and demagnetized by an external field.

硬/软铁磁体

(a) 硬铁磁材料的磁滞曲线，其对应的剩余磁化强度较大。这类材料不易被外磁场去磁化，故而被用作永久磁体。(b) 软铁磁材料的磁滞曲线，其对应的剩余磁化强度较小。这类材料容易在外场作用下被磁化或去磁化。

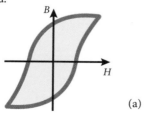

(a)

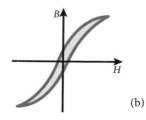

(b)

DIAMAGNETISM and PARAMAGNETISM
抗磁性和顺磁性

TERMS	术语
substance	物质
paramagnetic substance	顺磁性物质
diamagnetic substance	抗磁性物质
atoms (ions) with permanent magnetic dipole moments	具有永久磁偶极矩的原子（离子）
dipole	偶极子
randomly oriented dipoles	随机取向的偶极子
dipoles line up with the field	和磁场方向相同的偶极子
dipole induced in direction opposite to the applied field	感应偶极子与外磁场方向相反
susceptibility	磁化率
positive susceptibility	正磁化率
small susceptibility	小磁化率
Curie's law	居里定律
Curie temperature	居里温度

LANGUAGE OF PHYSICS
物理学用语

A **paramagnet** is a substance with permanent magnetic dipole moments that form an internal, induced magnetic field in the direction of the applied magnetic field.

顺磁体：一种具有永磁偶极矩的物质，其内感应磁场与外磁场同向。

A **diamagnetic material** is a substance that forms induced magnetic fields in the direction opposite to the applied magnetic field. Atoms of these materials have no permanent magnetic dipole moments.

抗磁材料：产生的感应磁场方向与外加磁场方向相反的一种物质。构成这些材料的原子没有永磁偶极矩。

Curie's law
The magnetization of a paramagnetic substance is proportional to the applied magnetic field and inversely proportional to the absolute temperature.

居里定律
顺磁物质的磁化强度与外加磁场成正比，与绝对温度成反比。

The **Curie temperature**, or Curie point, is the temperature above which certain materials lose their permanent magnetic properties (ferromagnets become paramagnetic).

居里温度（居里点）：高于此温度时某些材料会失去永久磁性（铁磁体呈现顺磁性）。

电磁学 **211**

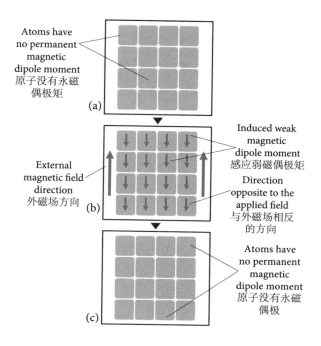

DIAMAGNETISM
Diamagnetic materials:
(a) normal;
(b) magnetic field applied;
(c) magnetic field removed.

抗磁性
抗磁性材料：
（a）常态时；
（b）施加外磁场时；
（c）去除外磁场时。

PARAMAGNETISM
Paramagnetic materials:
(a) normal;
(b) magnetic field applied;
(c) magnetic field removed.

顺磁性
顺磁性材料：
（a）常态时；
（b）施加外磁场时；
（c）去除外磁场时。

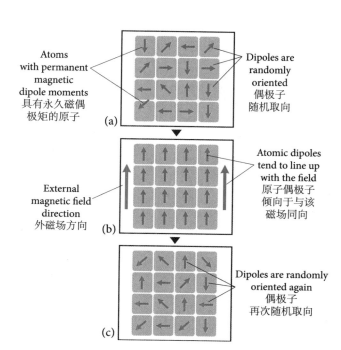

MAGNETIC FIELD of the EARTH
地磁场

TERMS 术语	
north pole	北极
south pole	南极
geographic pole	地理磁极
magnetic pole	磁极
Earth's magnetic field	地球磁场
Earth's nonuniform magnetic field	地球的非均匀磁场
Earth's magnetic field pattern	地球的磁场图谱
Earth's magnetic field source	地球的磁场源
reversal of the magnetic field direction	磁场方向的翻转
Van Allen radiation belt	范艾伦辐射带
cosmic ray	宇宙射线
Northern Lights	北极光
charged particle trapped (deflected) by the Earth's magnetic field	被地球磁场困住（偏转）的带电粒子
charged particles spiral around the field lines	带电粒子围绕磁场线旋转
charged particles originate from the Sun	带电粒子来自太阳
charged particles collide with atoms	带电粒子和原子发生碰撞
visible light	可见光

LANGUAGE OF PHYSICS 物理学用语

The **magnetic field of the Earth** is the magnetic field that extends from the Earth's interior out into space, where it meets the solar wind, a stream of charged particles emanating from the Sun. The magnetic field is generated by electric currents due to the motion of convection currents of molten iron in the Earth's outer core driven by heat escaping from the core.

地磁场：从地球内部延伸到太空的磁场，它在太空中与太阳释放出的带电粒子即太阳风相遇。热量从地球外核中向外逸会使地球外核中熔融态的铁做对流运动并产生电流，从而引发地磁场。

The **Northern Lights** occur when cosmic rays, electrically charged particles originating mainly from the Sun, become trapped in the Earth's atmosphere over the Earth's magnetic poles and collide with other atoms, causing them to emit visible lights.

当宇宙射线（主要是来自太阳的带电粒子）被困在地球磁极上方的大气层中，并与其他原子发生碰撞时，会导致原子受激发出可见光，即**北极光**。

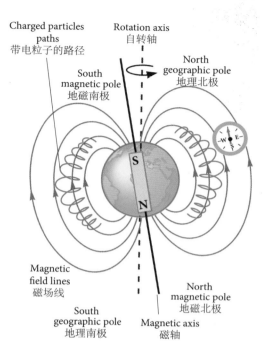

THE EARTH'S MAGNETIC FIELD LINES

A south magnetic pole is located at the north geographic pole and a north magnetic pole is located at the south geographic pole. The direction of the field has reversed several times during the last million years.

The Van Allen belts are made up of charged particles (electrons and protons) trapped by the Earth's nonuniform magnetic field.

地球磁场线

地磁南极位于地理北极附近,地磁北极位于地理南极附近。在过去的一百万年里,磁场方向已经发生过多次翻转。

范艾伦带由非均匀地磁场捕获的带电粒子(电子和质子)组成。

The Northern Lights 北极光

FARADAY'S LAW of ELECTROMAGNETIC INDUCTION
法拉第电磁感应定律

TERMS	术语
electromagnetic induction	电磁感应
loop of wire	线圈
loop of wire of a given area	给定面积的线圈
galvanometer	电流计
galvanometer needle	电流计指针
deflection in one direction	向一个方向偏转
magnet moves towards the loop	磁体靠近线圈
magnet moves away from the loop	磁体远离线圈
magnet holds stationary relative to the loop	磁铁相对于线圈保持静止
induced current	感应电流
set up of induced current	感应电流的产生
induced emf	感应电动势
change of magnetic flux	磁通量变化
magnetic flux threads the circuit	磁通量穿过电路
closed circuit	闭合电路
coil	线圈
area vector	面积矢量

LANGUAGE OF PHYSICS 物理学用语

Faraday's law of electromagnetic induction   **法拉第电磁感应定律**

The emf induced in a circuit is directly proportional to the time rate of change of magnetic flux through the circuit.

电路中感应的电动势与通过电路的磁通量的时间变化率成正比。

The **magnetic flux can be changed** by changing the magnetic field, the area of the loop, or the direction between the magnetic field and the area vector.

磁通量可以通过改变磁场、线圈面积或者磁场相对于面积矢量的方向来**改变**。

INDUCED CURRENT

When a magnet is moved toward a loop of wire connected to a galvanometer, the galvanometer's needle deflects as shown in (a). This shows that a current is induced in the loop. When the magnet is moved away from the loop in (c), the galvanometer deflects in the opposite direction, indicating that the induced current is opposite to that shown in (a).

感应电流

如图（a）所示，当磁体靠近线圈时，连接在线圈上的电流计的指针会发生偏转，这说明在线圈内感应出了电流。如图（c）所示，当磁体远离线圈时，电流计偏转到相反方向，说明感应电流与图（a）所示方向相反。

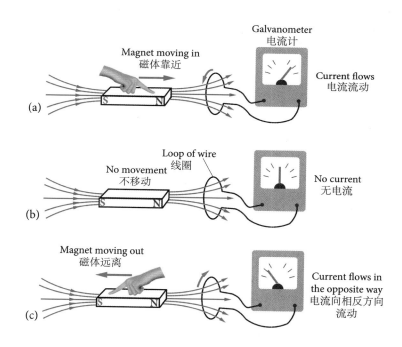

FARADAY'S LAW

The general statement that summarizes such experiments involving induced currents and emfs is known as Faraday's law of induction.

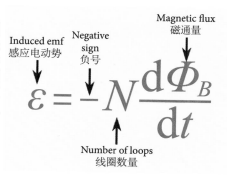

$$\varepsilon = -N \frac{d\Phi_B}{dt}$$

法拉第定律

法拉第定律概括总结了与感应电流和感应电动势相关的实验规律。

LENZ'S LAW
楞次定律

TERMS 术语

direction of induced emf	感应电动势方向
direction of induced current	感应电流方向
direction toward loop	靠近线圈的方向
external magnetic flux	外部磁通量
own magnetic flux	自行产生的磁通量
magnetic flux produced by current	电流产生的磁通量
magnetic flux counteracts the increase of current	磁通量抵消电流的增加
magnetic flux through loop	通过线圈的磁通量
loop	线圈
stationary loop	静止线圈
conducting loop	导线圈
changing magnetic field	变化磁场
polarity	极性

LANQUAGE OF PHYSICS 物理学用语

Lenz's law
The polarity of the induced emf is such that it tends to produce a current that will create a magnetic flux to oppose the change in magnetic flux through the loop.

楞次定律
感应电动势的极性是这样的，它倾向于产生感应电流。感应电流产生的磁通量总是要抵消线圈中原本磁通量的变化。

Lenz's law is a statement of the law of conservation of energy. Induced current tends to keep the original flux through the circuit from changing.

楞次定律是能量守恒定律的一种表述。感应电流总是要使电路中的初始磁通量保持不变。

MAGNETIC FLUX PRODUCED BY EXTERNAL MAGNETIC FIELD

When a magnet is moved toward the stationary conducting loop, a current is induced in the direction shown.

外磁场产生的磁通量

磁体靠近静止导线圈时,线圈中会感应出如图所示的电流。

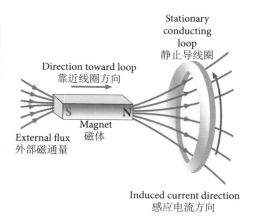

MAGNETIC FLUX PRODUCED BY INDUCED CURRENT

This induced current produces its own flux to the left to counteract the increasing external flux to the right.

感应电流产生的磁通量

感应电流产生向左的通量来抵消外部增加的向右的通量。

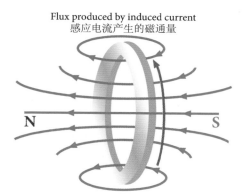

FARADAY'S EXPERIMENT

Experimental work conducted by M. Faraday in 1831 showed that an electric current could be induced in a circuit by a changing magnetic field.

法拉第实验

法拉第在1831年进行的实验表明,变化的磁场可以在电路中感生电流。

GENERATOR and MOTOR
发电机和电动机

TERMS 术语

generator	发电机
alternating current generator (AC generator)	交流发电机
direct current generator (DC generator)	直流发电机
commercial generator	商用发电机
frequency of generator	发电机频率
alternating emf	交变电动势
alternating emf varies sinusoidally with time	交变电动势随时间呈正弦变化
DC current	直流电流
pulsating DC current	脉冲直流电流
steady current	稳恒电流
slip ring	滑环
carbon brush	碳刷
axis of rotation	旋转轴
magnetic pole	磁极
wire loop	线圈
commutator	整流器

LANGUAGE OF PHYSICS 物理学用语

The **alternating current generator (AC generator)** is a device that converts mechanical energy to electrical energy. The AC generator is a source of alternating current. In the AC generator a coil of wire is manually rotated in an external magnetic field. Because the magnetic flux through the coil changes with time, the rotating coil has an alternating, sinusoidally varying emf and current induced in the coil.

交流发电机：把机械能转化为电能的装置。交流发电机是交流电电源。在外磁场中手动旋转交流发电机的线圈，通过线圈的磁通量将随时间变化，旋转线圈中则会产生交变的、正弦变化的电动势和感应电流。

A **motor** is device that convert electrical energy into mechanical energy.

电动机：把电能转化为机械能的装置。

A **motor** is a generator operating in reverse.

电动机就是反向运转的发电机。

AC GENERATOR

(a) Schematic diagram of an AC generator.
(b) The alternating emf induced in the loop plotted versus time.

交流发电机

（a）交流发电机原理图。
（b）线圈中感应交变电动势随时间的变化。

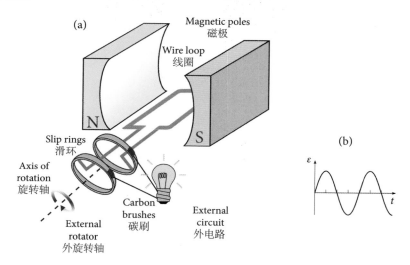

DC GENERATOR

(a) Schematic diagram of a DC generator.
(b) The emf plotted versus time fluctuates in magnitude but always has the same polarity.

直流发电机

（a）直流发电机原理图。
（b）电动势随时间变化图，电动势大小随时间变化，但极性不变。

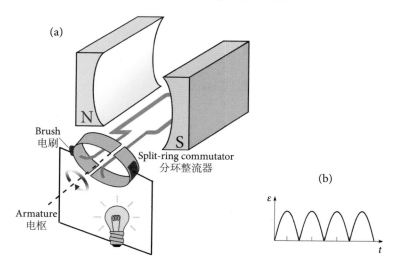

EDDY CURRENT
涡流

TERMS	术语
current	电流
circulating current	环流
eddy current	涡流
induced current	感应电流
current decreases steadily	电流稳步下降
Foucault current	傅科电流
Faraday's law of electromagnetic induction	法拉第电磁感应定律
changing flux	变化的磁通量
induced emf	感应电动势
bulk metal	块体金属
moving through a magnetic field	在磁场中运动
current clockwise	顺时针电流
current counterclockwise	逆时针电流
pivot	枢轴
braking system	制动系统
drag force	阻力
train slows down	列车减速

LANGUAGE OF PHYSICS 物理学用语

An **eddy current** is a localized electric current induced in a conductor by a varying magnetic field, due to Faraday's law of electromagnetic induction.

根据法拉第电磁感应定律,变化的磁场会在导体中激发感应电流,局域化的感应电流则被称为**涡流**。

Eddy currents flow in closed loops within conductors, in planes perpendicular to the magnetic field.

涡流沿导体内闭合回路流动,且闭合回路所在平面垂直于磁场。

Eddy currents are also called Foucault currents.

涡流也被称为傅科电流。

EDDY CURRENT FORMATION

The formation of eddy currents in a conductor moving through a magnetic field. As the plate enters or leaves the field, the changing flux sets up an induced emf, which causes eddy currents in the plate. This will cause the plate to slow down and eventually come to a stop.

涡流的形成

导体在磁场中运动，产生涡流。当平板进入或离开磁场时，变化的磁通量会产生感应电动势，从而在平板中产生涡流。这将导致平板减速并最终停止运动。

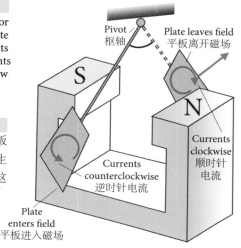

The braking systems on many subways make use of eddy currents. An electromagnet attached to the train is positioned near the steel rails. The braking action occurs when a large current is passed through the electromagnet. The relative motion of the magnet and rails induces eddy currents in the rails, and the direction of these currents produces a drag force on the moving train. Because the eddy currents decrease steadily in magnitude as the train slows down, the braking effect is quite smooth.

许多地铁的制动系统都利用了涡流。连接在列车上的电磁铁位于钢轨附近。当大电流通过电磁铁时，制动作用就发生了。磁体和轨道发生相对运动，并在阻碍列车移动的方向上产生涡流。由于列车减速时涡流的大小稳步降低，所以制动效果非常平稳。

MAXWELL'S EQUATIONS (Ⅰ)
麦克斯韦方程组（Ⅰ）

TERMS 术语

Gauss' law	高斯定律
total electric flux	总电通量
net electric flux	净电通量
electric field vector	电场矢量
closed surface	封闭曲面
closed volume	封闭体积
integral over a closed surface	封闭曲面的积分
net charge	净电荷
charge distribution	电荷分布
permittivity of free space	真空介电常数
Gauss' law in magnetism	磁场的高斯定律
net magnetic flux	净磁通量
magnetic field vector	磁场矢量
magnetic monopole	磁单极子

LANGUAGE OF PHYSICS 物理学用语

Gauss' law
The total electric flux through any closed surface equals the net charge inside that surface divided by ε_0.

高斯定律
通过任意封闭曲面的总电通量等于曲面内的净电荷除以 ε_0。

Gauss' law in magnetism
The net magnetic flux through a closed surface is zero.

磁场的高斯定律
通过封闭曲面的净磁通量为零。

GAUSS' LAW

This equation is Gauss' law. This law relates an electric field to the charge distribution that creates it.

高斯定律

这个方程就是高斯定律。该定律将电场与产生该电场的电荷分布联系起来。

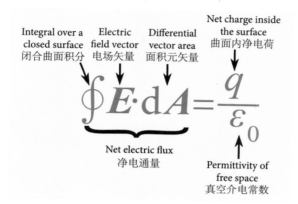

GAUSS' LAW IN MAGNETISM

This equation, which can be considered Gauss' law in magnetism, states that the number of magnetic field lines that enter a closed volume must equal the number that leave that volume. This implies that magnetic field lines cannot begin or end at any point. The fact that isolated magnetic monopoles have not been observed in nature can be taken as confirmation of this equation.

磁场的高斯定律

这个方程可以被看作磁场的高斯定律。它表明进入一个封闭体积的磁场线数目必须等于离开该体积的磁场线数目，即磁场线不能在任何一点开始或结束。在自然界中还没有观测到孤立的磁单极子，这一事实可作为对这一方程的实证。

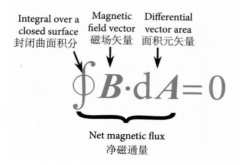

MAXWELL'S EQUATIONS (II)
麦克斯韦方程组（II）

TERMS 术语

Faraday's law of electromagnetic induction	法拉第电磁感应定律
line integral	线积分
line integral over a path	路径线积分
closed path	封闭路径
electric field vector	电场矢量
electric flux	电通量
magnetic flux	磁通量
changing magnetic flux	变化的磁通量
emf (electromotive force)	电动势
surface area bounded by a path	路径所围面积
Ampère-Maxwell law	安培-麦克斯韦定律
permeability of free space	真空磁导率
permittivity of free space	真空介电常数
rate of change of flux	通量变化率
conduction current	传导电流

LANGUAGE OF PHYSICS 物理学用语

Faraday's law of electromagnetic induction
The emf, which is the line integral of the electric field around any closed path, equals the rate of change of magnetic flux through any surface area bounded by that path.

法拉第电磁感应定律
电动势是电场沿任意闭合路径的线积分，它等于通过该闭合路径所围面积的磁通量变化率。

Ampère-Maxwell law
The line integral of the magnetic field around any closed path is the sum of the net conduction current through that path and the rate of change of electric flux through any surface bounded by that path.

安培-麦克斯韦定律
磁场沿任意闭合路径的线积分等于流过该回路的净电流与通过该闭合路径所围面积的电通量变化率之和。

FARADAY'S LAW OF INDUCTION

This equation summarizes Faraday's law of induction. This law describes how an electric field can be induced by a changing magnetic flux.

法拉第电磁感应定律

这个方程总结了法拉第电磁感应定律。该定律描述了变化的磁通量如何产生电场。

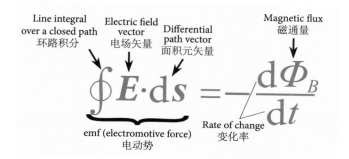

AMPÈRE-MAXWELL LAW

This equation is usually called the Ampère-Maxwell law, and is the generalized form of Ampère's law, and describes the creation of a magnetic field by a changing electric flux and electric currents.

安培-麦克斯韦定律

这个方程通常被称为安培-麦克斯韦定律，是安培定律的广义形式，描述了变化的电通量和电流如何产生磁场。

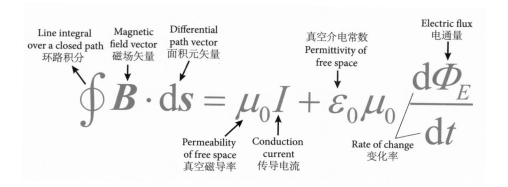

INDUCTANCE
电感

TERMS	术语
self-induction	自感
self-induced emf	自感电动势
inductance of coil	线圈的电感
N-turn coil	N匝线圈
magnetic flux through loop	通过线圈的磁通量
magnetic flux changes in time	磁通量随时间变化
magnetic flux induces emf	磁通量产生电动势
loop	回路
switch	开关
battery	电池
equilibrium value of current	电流平衡值
mutual inductance	互感系数
mutual induction	互感
galvanometer	电流计

LANGUAGE OF PHYSICS   物理学用语

Inductance is a measure of the opposition to any change in current. The inductance of a device depends on its geometry (size, shape of coil) or material characteristics.

电感：用于衡量电流变化的阻力。器件的电感取决于其几何形状（线圈的大小、形状）或材料特性。

Self-induction is the production of an electromotive force (emf) in a circuit when the current in that circuit is varied. The self-induced emf is always proportional to the time rate of change of the current.

自感：电路中电流变化时产生电动势（emf）的现象。自感电动势总是与电流随时间的变化率成正比。

Mutual induction is the production of an electromotive force in a circuit by a change in the current in an adjacent circuit which is linked to the first by the flux lines of a magnetic field.

互感：一个电路通过磁场的磁通量线与另一个相邻电路连接起来，相邻电路中的电流发生变化时会在该电路中产生电动势。

SELF-INDUCTION

After the switch is closed, the current produces a magnetic flux through the loop. As the current increases toward its equilibrium value, the flux changes in time and induces an emf in the loop. The battery drawn with dashed lines represents the self-induced emf.

自感

开关闭合后，电流会在回路产生磁通量。在电流增大至其平衡值的过程中，磁通量随时间变化，并在回路中产生电动势。用虚线画出的电池表示自感电动势。

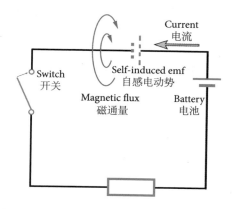

MUTUAL INDUCTION

A current in coil 1 sets up a magnetic flux, part of which passes through coil 2. The emf induced by mutual induction in the coil 2 is proportional to the rate of current change in coil 1.

互感

线圈1中的电流产生磁通量，其中一部分通过线圈2。线圈2中互感产生的电动势与线圈1中电流变化的速率成正比。

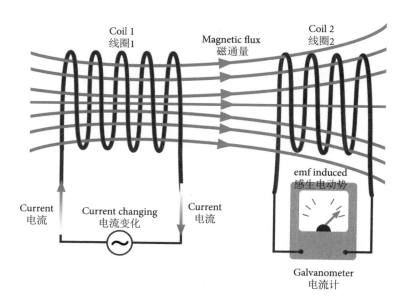

OSCILLATION in an LC CIRCUIT
LC电路中的振荡

TERMS 术语

inductor	电感器
capacitor	电容器
charged capacitor	带电电容器
fully charged capacitor	充满电的电容器
discharged capacitor	放电电容器
electric field of capacitor	电容器的电场
capacitor connected to inductor	连接电感器的电容器
magnetic field of inductor	电感器的磁场
switch closed	开关闭合
LC circuit	LC电路
resistanceless LC circuit	无电阻的LC电路
energy transfer	能量转换
energy stored in capacitor	电容器储存的能量
energy stored in inductor	电感器储存的能量
energy dissipated as Joule heat	以焦耳热形式耗散的能量
current oscillation	电流振荡
charge oscillation	电荷振荡
energy oscillation	能量振荡
initial charge	起始电荷
angular frequency of oscillation	振荡角频率
inductance	电感
capacitance	电容
current 90° out of phase with stored charge	电流与储存的电荷差90° 相位
graph of charge versus time	电荷与时间的关系图
current versus time	电流与时间
extreme value of charge	电荷极值
extreme value of current	电流极值

LANGUAGE OF PHYSICS 物理学用语

An **LC circuit** is a circuit containing an inductor and a capacitor in series. The resistance of the circuit is zero.

LC电路：由一个电感器和一个电容器串联而成的电路。LC电路的电阻为零。

The current through, and the charge on, the capacitor oscillate in the LC circuit. Due to there being zero resistance no energy is dissipated as Joule heat and the oscillations persist.

LC电路中的电流和电容器两端的电荷发生振荡。由于电阻为零，能量不会以焦耳热形式耗散，振荡持续。

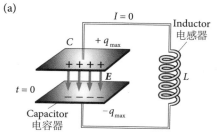
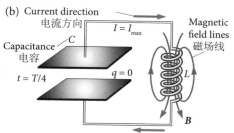
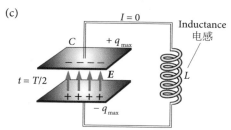
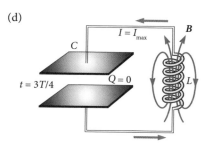
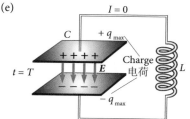

ENERGY TRANSFER IN A RESISTANCELESS LC CIRCUIT

(a) The capacitor is fully charged at $t = 0$. All of the energy in the circuit is stored as potential energy in the capacitor. The current I is zero. No energy is stored in the inductor.

(b) The capacitor begins to discharge and the energy stored in its electric field decreases. At the same time, the current increases and some energy is now stored in the magnetic field of the inductor. When the capacitor is fully discharged, it stores no energy. At this time, the current reaches its maximum value I_{max} and all of the energy is stored as magnetic energy in the inductor.

(c), (d) The process then repeats in the reverse direction.

(e) At time $t = T$ the system returns to the initial position corresponding to $t = 0$.

无电阻LC电路中的能量转换

（a） $t = 0$时电容器充满电。电路中所有能量都以势能的形式存储在电容器中。电流I是0。电感器没有能量。

（b）电容器开始放电，其电场中储存的能量减少。同时，电流增加，此时一些能量储存在电感器的磁场中。电容器完全放电后，不再储存能量。此时，电流达到最大值I_{max}，所有能量以磁能的形式储存在电感器中。

（c）、（d） 随后逆向重复这一过程。

（e） $t = T$时，系统恢复到与$t = 0$时一样的状态。

RC, RL and RLC CIRCUITS
RC电路、RL电路和RLC电路

TERMS 术语

circuit	电路
steady-state circuit	稳态电路
open circuit	开路
capacitor	电容器
uncharged capacitor	未充电的电容器
fully charged capacitor	充满电的电容器
charging of capacitor	电容器充电
discharging of capacitor	电容器放电
switch	开关
time constant	时间常数
resistor	电阻器
current	电流
varying current	变化的电流
current versus time	电流与时间
current rises to the equilibrium value	电流增大至平衡值
current decays exponentially	电流呈指数衰减
current oscillates sinusoidally	电流正弦振荡
series RL circuit	RL串联电路
series RLC circuit	RLC串联电路
inductor	电感器
emf of battery	电池电动势
time rate of change of current	电流随时间的变化率
energy stored in capacitor	电容器储存的能量
energy stored in inductor	电感器储存的能量
damped simple harmonic motion	有阻尼的简谐运动
critical resistance	临界电阻

LANGUAGE OF PHYSICS 物理学用语

An **RC series circuit** is a circuit containing a resistor and a capacitor in series.

RC串联电路：由一个电阻器和一个电容器串联而成的电路。

An **RL series circuit** is a circuit containing a resistor and an inductor in series.

RL串联电路：由一个电阻器和一个电感器串联而成的电路。

An **RLC series circuit** is a circuit containing a resistor, an inductor and a capacitor in series.

RLC串联电路：由一个电阻器、一个电感器和一个电容器串联而成的电路。

RC CIRCUIT

A charged capacitor connected to a resistor. After the switch is closed, a varying current is set up in the direction shown and the charge on the capacitor decreases exponentially with time. (a) Plot of capacitor charge versus time. (b) Plot of current versus time.

RC电路

带电电容器与电阻器相连。开关闭合后，电路中产生如图所示的变化电流，电容器上的电荷随时间呈指数下降。(a) 电容器电荷与时间的关系图。(b) 电流与时间的关系图。

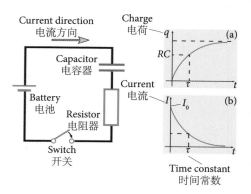

RL CIRCUIT

A series RL circuit. As the current increases toward its maximum value, the inductor produces an emf that opposes the increasing current. (a) Plot of the current versus time. (b) Plot of the time rate of change of current versus time.

RL电路

RL串联电路。随着电流增大至最大值，电感器产生电动势来阻止电流增加。(a) 电流与时间的关系图。(b) 电流随时间的变化率与时间的关系图。

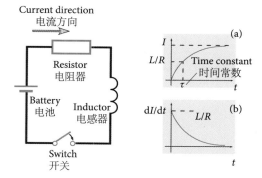

RLC CIRCUIT

A series RLC circuit. Charge versus time for a damped RLC circuit.

RLC电路

RCL串联电路。RCL阻尼电路中电荷与时间的关系图。

RESISTOR in an AC CIRCUIT
交流电路中的电阻器

TERMS 术语	
alternating current	交流电流
AC circuit	交流电路
AC generator	交流发电机
closed loop	闭合回路
resistor	电阻器
voltage drop	电压降
instantaneous voltage drop	瞬时电压降
voltage drop across resistor	电阻两端的电压降
phasor diagram	相量图
phasor	相量
average value of current	电流平均值
maximum current	最大电流
RMS current	均方根电流
RMS voltage	均方根电压
current and voltage in phase	电流和电压同相位
current and voltage vary sinusoidally	呈正弦变化的电流和电压

LANGUAGE OF PHYSICS 物理学用语

In an **AC circuit consisting of a generator and a resistor**, the current in the circuit is in phase with the voltage. The current and voltage reach their peak values at the same time.

在**由发电机和电阻器组成的交流电路**中，电流与电压同相位。电流和电压同时达到峰值。

The **RMS value** or **effective current** of an AC current is the equivalent DC current that generates the same heat in the circuit. The effective value of an alternating current is equal to 70.7% of the maximum or peak value of the AC current. An ammeter measures the effective current in the circuit.

交流电流的**均方根值（RMS）或有效电流**：在电路中产生相同热量的等效直流电流。交流电流的有效值等于交流电流最大值或峰值的70.7%。安培表可测量电路中的有效电流。

The **RMS voltage** or **effective voltage** in an AC circuit is a constant value of the voltage that produces the same effect as the alternating voltage. The effective value of an alternating voltage is equal to 70.7% of the maximum or peak value of the AC voltage. A voltmeter measures the effective voltage in the circuit.

交流电路中的**电压均方根值（RMS）或有效电压**：产生与交流电压相同效果的稳恒电压值。交流电压的有效值等于交流电压最大值或峰值的70.7%。电压表可测量电路中的有效电压。

RESISTIVE CIRCUIT

A circuit consisting of a resistor R connected to an AC generator.

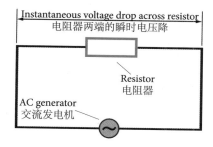

电阻电路

该电路由一个电阻器R与交流发电机连接而成。

CURRENT AND VOLTAGE

Plots of the current and voltage across the resistor as a function of time. The current is in phase with the voltage. It means that the voltage is zero when the current is zero, maximum when the current is maximum and minimum when the current is minimum.

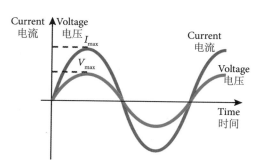

电流和电压

通过电阻器的电流和电阻器两端电压随时间变化的曲线图。电流与电压同相位，即电流为零时电压为零，电流最大时电压最大，电流最小时电压最小。

PHASOR DIAGRAM

A phasor diagram for the resistive circuit, showing that the current is in phase with the voltage.

相量图

电阻电路的相量图，表明电流与电压同相位。

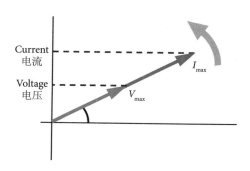

INDUCTOR in an AC CIRCUIT
交流电路中的电感器

TERMS 术语

AC circuit consisting of an inductor	连接一个电感器的交流电路
inductive circuit	电感电路
emf induced in an inductor	电感器中产生的电势
inductor connected to terminals of AC generator	连接在交流发电机两端的电感器
current and voltage out of phase	电流和电压不同相位
current lags voltage by 90°	电流滞后电压90°
inductive reactance	感抗
instantaneous voltage drop across inductor	电感器两端瞬时电压降
phasor diagram	相量图
current (voltage) phasor	电流（电压）相量
maximum value of current (voltage)	电流（电压）最大值
oscillation period	振荡周期

LANGUAGE OF PHYSICS 物理学用语

In an **AC circuit consisting of a generator and an inductor**, the current lags behind the voltage by 90°.

在一个发电机和一个电感器组成的交流电路中，电流滞后电压90°。

The voltage reaches its maximum value one quarter of a period before the current reaches its maximum value.

电流比电压晚1/4个周期达到最大值。

Inductive reactance tends to impede changes in the flow of charges in an AC circuit. The inductive reactance is proportional to the frequency of the AC source. A low-frequency AC source causes a low reactance, whereas a high-frequency source causes a high reactance.

感抗总是阻碍交流电路中电荷流的变化。感抗的大小与交流电源的频率大小成正比。低频交流电产生低电抗，而高频交流电产生高电抗。

INDUCTIVE CIRCUIT

A circuit consisting of an inductor L connected to an AC generator.

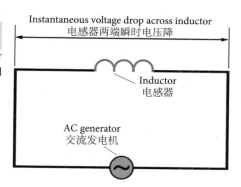

电感电路

该电路由一个电感器L和一个交流发电机连接而成。

CURRENT AND VOLTAGE

Plots of the current and voltage across the inductor as a function of time. The current lags the voltage by 90°.

电流和电压

电感器上电流和电压随时间变化的曲线图。电流相位滞后电压90°。

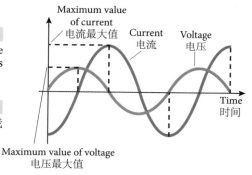

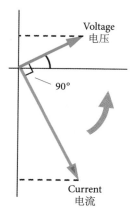

PHASOR DIAGRAM

The phasor diagram for the inductive circuit. The angle between the current phasor and voltage phasor is 90°.

相量图

电感电路相量图。电流相量和电压相量之间的夹角为90°。

CAPACITOR in an AC CIRCUIT
交流电路中的电容器

TERMS	术语
AC circuit consisting of a capacitor	连接一个电容器的交流电路
capacitor connected across terminals of AC generator	连接在交流发电机两端的电容器
current with voltage not in phase	电流和电压不同相位
current with voltage 90° out of phase	电流与电压相差90°相位
instantaneous voltage drop across capacitor	电容器两端瞬时电压降
phasor diagram	相量图
current leads voltage by 90°	电流超前电压90°
capacitive reactance	容抗
current (voltage) phasor	电流（电压）相量

LANGUAGE OF PHYSICS　　物理学用语

In an **AC circuit consisting of a generator and a capacitor**, the current leads the voltage by 90°.

在一个发电机和一个电容器组成的交流电路中，电流超前电压90°。

The current reaches its maximum value one quarter of a period before the voltage reaches its maximum value.

电流比电压提前1/4个周期达到最大值。

Capacitive reactance tends to impede the buildup of charge in an AC circuit. It is the capacitive analogue to resistance in an AC circuit. The capacitive reactance is inversely proportional to the frequency of the AC source. A low-frequency AC source causes a high reactance, whereas a high-frequency source causes a low reactance.

容抗总是阻碍交流电路中电荷的累积。交流电路中，电容与电阻类似。容抗与交流电源的频率成反比，低频交流电源产生高电抗，而高频交流电源产生低电抗。

CAPACITIVE CIRCUIT

A circuit consisting of a capacitor C connected to an AC generator.

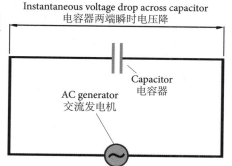

电容电路

一个电容器C与一个交流发电机连接而成的电路。

CURRENT AND VOLTAGE

Plots of the current and voltage across the capacitor as a function of time. The voltage lags behind the current by 90°.

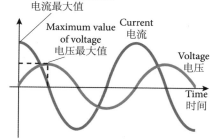

电流和电压

电容器上的电流和电压随时间变化的曲线图。电压滞后电流90°。

PHASOR DIAGRAM

Phasor diagram for the purely capacitive circuit. Projections of the phasors onto the vertical axis give the instantaneous values of voltage and current.

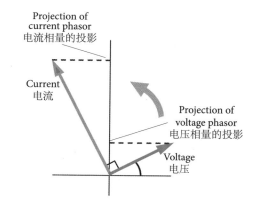

相量图

纯电容电路的相量图。相量在垂直轴上的投影表示电压和电流的瞬时值。

RLC SERIES CIRCUIT
RLC串联电路

TERMS 术语

applied voltage varies sinusoidally	外加电压呈正弦变化
phase angle between current and voltage	电流与电压间的相位角
voltage across resistor in phase with current	电阻器的电压与电流同相位
voltage across inductor leads the current by 90°	电感器的电压超前电流90°
voltage across capacitor lags behind the current by 90°	电容器的电压滞后电流90°
impedance	阻抗
power factor	功率因数
average power	平均功率
resonance frequency	共振频率
quality factor	品质因子
high-Q (low-Q) circuit	高品质（低品质）电路

LANGUAGE OF PHYSICS 物理学用语

Impedance is composed of the resistance, inductive reactance and the capacitive reactance of an AC circuit. It is a measure of the opposition to the flow of charges in an AC circuit. It is the AC analogue to resistance in a DC circuit.

阻抗：由交流电路中的电阻、感抗以及容抗组成。它是交流电路中电荷流动阻力的量度，类似于直流电路中的电阻。

Resonance occurs when the inductive reactance is equal to the capacitive reactance and then the maximum current is obtained in an AC circuit.

共振：当感抗等于容抗时，就会发生共振，此时交流电路中的电流达到最大值。

Resonant frequency is defined as a frequency of an AC circuit that causes resonance in the circuit. It depends on the inductance and capacitance of the circuit.

共振频率：交流电路中能够引起电路发生共振的频率。它取决于电路中的电感和电容。

The **power factor** is the ratio of the power consumed in an AC circuit to the power applied to the circuit.

功率因数：交流电路中消耗的功率与施加于电路上的功率之比。

RLC SERIES CIRCUIT

A series circuit consisting of a resistor, an inductor, and a capacitor connected to an AC generator.

RLC串联电路

由一个电阻器、一个电感器和一个电容器与一个交流发电机串联而成的电路。

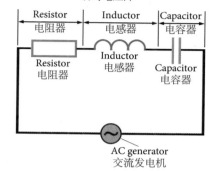

PHASE RELATIONSHIPS

Phase relationships in the series RLC circuit.

相位关系

RLC串联电路中的相位关系。

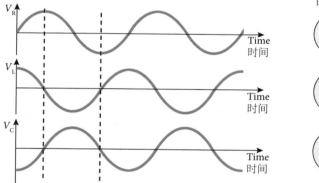

Phase relationships between the peak voltage and current phasor for

以下元件中电压和电流峰值相位之间的相位关系

Resistor 电阻器

Inductor 电感器

Capacitor 电容器

PHASOR DIAGRAM

The phasor diagram for the series RLC circuit. The phasor V_R is in phase with the current phasor I_{max}, the phasor V_L leads I_{max} by 90°, and the phasor V_C lags I_{max} by 90°. The total voltage V_{max} makes an angle ϕ with I_{max}.

相量图

RLC串联电路的相量图。相量V_R与电流相量I_{max}同相位，相量V_L超前I_{max} 90°，相量V_C落后I_{max} 90°。总电压V_{max}与I_{max}之间夹角为ϕ。

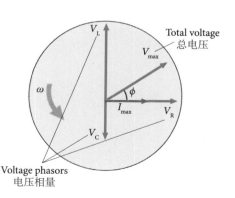

TRANSFORMER
变压器

TERMS 术语

transformer	变压器
step-up transformer	升压变压器
step-down transformer	降压变压器
coil of wire wound around a core	绕在铁芯上的线圈
core of soft iron	软铁芯
primary winding (turns)	初级绕组
left coil connected to input AC voltage source	左边线圈与交流输入电压源相连
secondary winding (turns)	次级绕组
right coil connected to load resistor	右边线圈与负载电阻相连
magnetic flux through one coil passes through the other coil	通过一个线圈的磁通量穿过另一个线圈
transmission line	输电线路
voltage stepped up	电压升高
voltage stepped down	电压降低
AC voltage	交流电压
output voltage	输出电压

LANGUAGE OF PHYSICS　　　　　物理学用语

A **transformer** is a device that can increase (or decrease) the AC voltage or current without causing appreciable changes in the power delivered.

变压器：一种能够增大（或减小）交流电压或电流且过程中不会导致输出功率发生明显变化的装置。

A **step-up transformer** is a transformer that increases voltage from the primary coil to the secondary coil.

升压变压器：使次级线圈电压高于初级线圈电压的变压器。

A **step-down transformer** is a transformer that decreases voltage from the primary coil to the secondary coil.

降压变压器：使次级线圈电压低于初级线圈电压的变压器。

IDEAL TRANSFORMER

An ideal transformer consists of two coils wound on the same soft iron core. An AC voltage is applied to the primary coil, and the output voltage appears across the load resistance.

理想变压器

理想变压器由两个线圈绕同一个软铁芯组成。在初级线圈上施加一个交流电压时，会在负载电阻上产生输出电压。

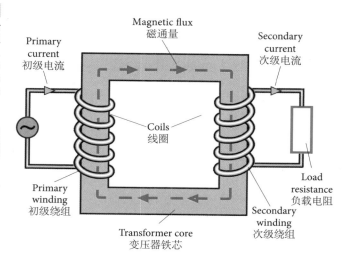

Transformers are useful for transmitting power over long distances. The voltage is stepped up to around 230,000 V at the power station, then stepped down to around 20,000 V at a distributing station, and finally stepped down to 120-220 V at the customer's electrical sockets.

变压器对远距离传输电力很有用。在发电站，电压被升高到约230 000 V，在配电站会降低到约20 000 V，最后到达用户的电源插座时降低到120~220 V。

ELECTROMAGNETIC WAVE
电磁波

TERMS 术语

electromagnetic wave	电磁波
plane electromagnetic wave	平面电磁波
transverse electromagnetic wave	横向电磁波
linearly polarized electromagnetic wave	线偏振电磁波
electromagnetic wave generated by oscillating electric charges	振荡电荷产生的电磁波
electromagnetic waves carry energy and momentum	电磁波具有能量和动能
electromagnetic waves exert pressure on a surface	电磁波对表面施加压力
electromagnetic waves predicted by Maxwell's equations	麦克斯韦方程组预言的电磁波
electromagnetic waves consist of oscillating electric and magnetic fields	电磁波由振荡电场和磁场组成
Poynting vector	坡印亭矢量

LANGUAGE OF PHYSICS 物理学用语

Electromagnetic waves are synchronized oscillations of electric and magnetic fields that propagate at the speed of light through a vacuum.

电磁波：由电场和磁场的同步振荡产生，并能在真空中以光速传播。

The electric wave and the magnetic wave are always perpendicular to each other and to the direction of wave propagation.

电磁波的电场分量和磁场分量不仅相互垂直而且垂直于波的传播方向。

The **intensity of radiation** is the total energy of an electromagnetic wave impinging on a unit area in a unit period of time. It is the power per unit area.

辐射强度：电磁波在单位时间作用在单位面积上的总能量，即单位面积的功率。

ELECTROMAGNETIC FIELD

A changing magnetic field produces an electric field. A changing electric field produces a magnetic field.

电磁场

变化的磁场产生电场，变化的电场产生磁场。

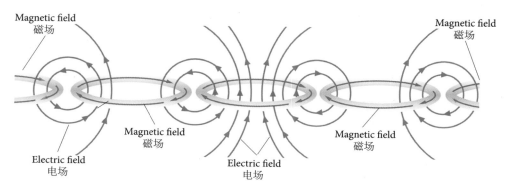

ELECTROMAGNETIC WAVE

Representation at one instant of a sinusoidal, plane-polarized electromagnetic wave moving in the positive *x* direction at the speed of light. The electric and magnetic field components are perpendicular to each other and also perpendicular to the direction of wave propagation.

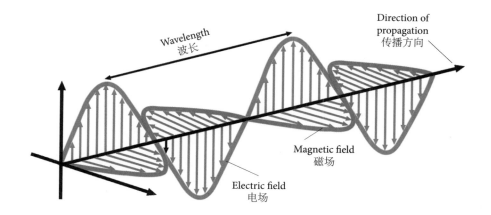

电磁波

图示为一列以光速沿 *x* 轴正方向传播的正弦平面极化电磁波某一瞬时示意图。电场和磁场分量相互垂直，同时二者也垂直于波的传播方向。

ELECTROMAGNETIC SPECTRUM
电磁波谱

TERMS	术语
electromagnetic wave type	电磁波的类型
wide range	宽频带
frequency	频率
wavelength	波长
form of radiation	辐射形式
radio wave	无线电波
microwave	微波
microwave generated by electronic device	电子器件产生的微波
infrared wave (IR) or heat wave	红外波或热波
IR generated by hot body and molecule	高温物体或高温分子产生的红外波
visible light	可见光
visible light produced by rearrangement of electrons in atoms and molecules	原子和分子内电子重新排列产生的可见光
ultraviolet light (UV)	紫外线
ultraviolet light generated by the Sun	太阳产生的紫外线
X-ray	X射线
X-ray generated by deceleration of high-energy electron	高能量电子减速产生的X射线
gamma ray	伽马射线
gamma ray emitted by radioactive nuclei during nuclear reaction	核反应中放射性原子核产生的伽马射线

LANGUAGE OF PHYSICS 物理学用语

The **electromagnetic spectrum** is the complete range of electromagnetic waves from the longest radio waves, through microwaves down to infrared rays, visible light, ultraviolet light, X-rays, and the shortest waves — the gamma rays.

电磁波谱：电磁波的完整范围，从最长的无线电波到微波、红外线、可见光、紫外线、X射线，再到最短的伽马射线。

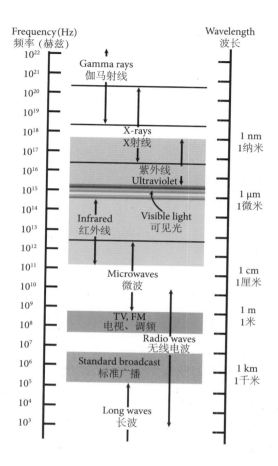

22-m radio telescope for mm and cm radio waves. Located at the foot of Mount Kishka near Simeiz (Crimea, Ukraine).

覆盖毫米和厘米级无线电波的22米射电望远镜。该望远镜位于西梅兹（Simeiz，位于乌克兰克里米亚）附近的基什卡山（Mount Kishka）山脚下。

OPTICS

A silica bead with a radius of 50 nanometers trapped levitating in a beam of light. Optical tweezers exploit the fact that light can exert pressure on matter. Although the radiation pressure even from an intense laser beam is quite small, the 2018 Noble Prize winner in Physics Ashkin was the first person to show that it was large enough to support a nanoparticle, countering gravity, effectively levitating it. The tweezers could trap microscale objects, as well as atoms and molecules, and manipulate viruses, living cells, and subcellular components.

半径为50纳米的硅珠悬浮在一束光中。光镊利用了光可以对物质施加压力这一特性。即使来自高强度激光束的辐射压力非常小，但2018年诺贝尔物理学奖得主阿什金首次证明这种压力足以支撑纳米粒子对抗重力，使其悬浮。光镊可以捕获微小的物体以及原子和分子，并操纵病毒、活细胞和亚细胞成分。

光学

OPTICS

Optics is the branch of physics which involves the behaviour and properties of light, including its interaction with matter and the construction of instruments that use or detect it.

光学

光学：物理学的一个分支，研究光的行为和性质，包括光与物质间的相互作用以及使用或探测光的仪器的构造。

BRANCHES OF OPTICS

Geometrical optics is the analysis of an optical system in terms of light rays that travel in straight lines.

Wave optics is the analysis of an optical system in terms of the wave nature of light.

Quantum optics is the study of light in terms of little bundles of electromagnetic energy, called photons.

光学分支

几何光学：从光线的直线传播角度对光学系统的分析。

波动光学：从光的波动本质的角度对光学系统的分析。

量子光学：从微小的电磁能量包即光子的角度对光的研究。

NATURE of LIGHT
光的本质

TERMS	术语
light	光
source of light	光源
light beam	光束
light ray	光线
light intensity	光强
corpuscular (particle) theory	微粒（粒子）理论
wave theory	波动理论
dual nature	二象性
particle-like behavior	粒子行为
high-frequency electromagnetic wave	高频电磁波
photon	光子
energy of photon	光子的能量
Planck's constant	普朗克常数
speed of light	光速
Roemer's method	罗默方法
Fizeau's technique	斐索（干涉）技术
geometric optics	几何光学
ray approximation	射线近似

LANGUAGE OF PHYSICS

物理学用语

In some cases (such as reflection, refraction, diffraction, polarization, interference) **light acts like a wave** and in others (such as the photoelectric effect and interaction of light with matter) **it acts like a particle**.

在某些情况（如反射、折射、衍射、偏振、干涉）下**光像波**，而在其他情况（如光电效应和光与物质的相互作用）下**光像粒子**。

A **photon** is a quantum of an electromagnetic field possessing a fixed energy. In terms of energy, each photon equals $h\nu$, where h is Planck's constant and ν is the frequency of the propagating electromagnetic wave.

光子：能量一定的电磁场的量子。每个光子的能量为$h\nu$，其中h为普朗克常数，ν为传播的电磁波频率。

The **wavelength** is the physical distance covered by one cycle of an electromagnetic wave in the form of a sinusoidal wave. It is inversely proportional to the frequency.

波长：一个正弦波形式的电磁波一个周期所覆盖的物理距离。它与频率成反比。

The **ray approximation** assumes that a wave travels through a medium in straight lines in the direction of the rays.

射线近似：假设波在介质中沿射线方向呈直线传播。

IS LIGHT A WAVE OR A PARTICLE?

Light is intimately involved with our daily lives. In physics, there are two theories by which light can be defined: the first theory defines light as particles and the second theory as waves.

When in a vacuum, for instance in space, where no matter is present, light travels directly in straight lines. However, when light comes into contact with water, air, and other matter — it may be "absorbed", "transmitted", "reflected", "scattered", or "polarized". The "transmitted", "reflected", or "scattered" light allows our eyes to see the color and shape of objects.

Being a wave, light can interfere with other light waves, canceling or amplifying them. Also light can be "dispersed" in media where the speed of the wave varies for different frequencies.

The color of light is determined by its frequency. Light in a vacuum travels at a speed of approximately 300,000 kilometers per second.

光是波还是粒子？

光与我们的日常生活密切相关。在物理学中，有两种定义光的理论：第一种理论将光定义为粒子，第二种理论将光定义为波。

在真空中，例如在没有物质的空间中，光直接沿直线传播。然而，当光与水、空气和其他物质接触时，它可能被"吸收""透射""反射""散射"或发生"偏振"。透射、反射或散射的光使我们的眼睛能够看到物体的颜色和形状。

作为一种波，光可以与其他光波相干涉，如抵消或放大光波。在介质中，不同频率的光速度可能不同，从而发生散射。

光的颜色由它的频率决定。光在真空中的传播速度大约是每秒30万千米。

REFLECTION of LIGHT
光的反射

TERMS	术语
light ray	光线
beam of light	光束
incident light	入射光
reflected light	反射光
surface	表面
flat surface	平坦表面
smooth surface	光滑表面
rough surface	粗糙表面
mirror-like surface	镜面
reflecting surface	反射面
medium	介质
boundary	边界
plane of incidence	入射面
incident ray	入射光
reflected ray	反射光
normal	法线
reflection	反射
specular reflection	镜面反射
diffuse reflection	漫反射
angle of incidence	入射角
angle of reflection	反射角
random direction	任意方向
same plane	同一平面

LANGUAGE OF PHYSICS 物理学用语

Law of reflection
The angle of reflection equals the angle of incidence. The incident ray, the normal, and the reflected ray all lie in the same plane.

反射定律
反射角等于入射角。入射光、法线和反射光都在同一平面上。

Angle of incidence is the angle that the incident ray makes with the normal, which is a line drawn perpendicular to the surface at the point where the incident ray strikes.

入射角：入射光与法线的夹角。法线是一条通过入射点并垂直于该点所在表面的直线。

Angle of reflection is the angle that the reflected ray makes with the normal, which is a line drawn perpendicular to the surface at the point where the incident ray strikes.

反射角：反射光与法线的夹角。法线是一条通过入射点并垂直于该点所在表面的直线。

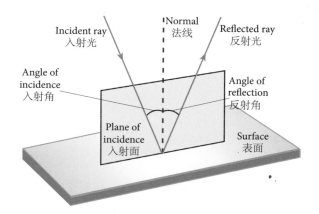

LAW OF REFLECTION
The incident ray, the reflected ray, and the normal all lie in the same plane. According to the law of reflection, the angle of incidence and the angle of reflection are equal.

反射定律
入射光、反射光和法线在同一平面内。根据反射定律，入射角与反射角相等。

This picture, taken in the Carpatian mountains, shows the reflection of landscape in the lake Synevyr.

这张照片摄于喀尔巴阡山，展示了库尼贡达湖景观的倒影。

SPECULAR REFLECTION
Schematic representation of specular reflection, where the reflected rays are all parallel to each other.

DIFFUSE REFLECTION
Schematic representation of diffuse reflection, where the reflected rays travel in random directions.

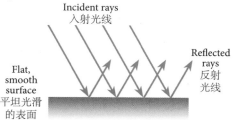
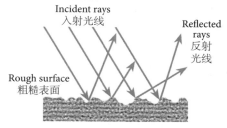

镜面反射
镜面反射示意图：反射光线彼此平行。

漫反射
漫反射示意图：反射光沿任意方向传播。

REFRACTION of LIGHT
光的折射

TERMS	术语
medium	介质
transparent medium	透明介质
light ray	光线
incident light ray	入射光线
reflected light ray	反射光线
refracted light ray	折射光线
light ray encounters a boundary	光线遇到边界
light ray bends at boundary	光线在边界处弯曲
light ray enters medium	光线进入介质
speed of light	光速
speed of light in a medium	介质中的光速
speed of light in a vacuum	真空中的光速
angle of incidence	入射角
angle of refraction	折射角
Snell's law*	斯涅尔定律
index of refraction	折射率
dimensionless number	无量纲数
greater than unity	比1大

LANGUAGE OF PHYSICS   物理学用语

Law of refraction

The incident ray, the reflected ray, the refracted ray, and the normal to the surface all lie in the same plane. The ratio of the sine of the angle of incidence to the sine of the angle of refraction is equal to the relative index of refraction. This is known as Snell's law.

折射定律

入射光、反射光、折射光和表面的法线都在同一平面上。入射角的正弦与折射角的正弦之比等于相对折射率。这就是斯涅尔定律。

The **angle of refraction** is the angle that the refracted ray makes with the normal, which is a line drawn perpendicularly to the surface at the point where the incident ray strikes.

折射角：折射光与法线的夹角。法线是一条过入射点并垂直于该点所在表面的直线。

The **relative index of refraction** is equal to the ratio of the speed of light in the first medium to the speed of light in the second medium.

相对折射率等于第一介质中的光速与第二介质中的光速之比。

* The law of refraction is known as Descartes' law in France and as Snell's law in English speaking countries.

* 折射定律在法国被称为笛卡儿定律，在英语国家被称为斯涅耳定律。

光学 **253**

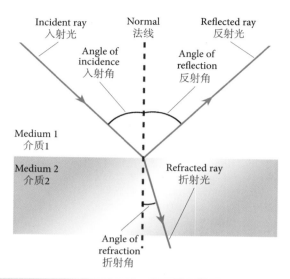

LAW OF REFRACTION

All rays and the normal lie in the same plane. The refracted ray is bent towards the normal because the speed of light in medium 2 is less than the speed of light in medium 1.

折射定律

所有光线和法线都在同一平面上。由于在介质2中的光速小于介质1中的光速，折射光线偏向法线。

BENDING OF LIGHT

Light incident on the glass block bends both when it enters the block and when it leaves the block. Because the sides of the block are parallel, the emerging ray is parallel to the original incoming ray, but shifted sideways a little. Note also the reflection at the inner surface of the block, and the refraction at the upper surface of the internally reflected ray.

光线弯曲

入射到玻璃块上的光线在进入和离开玻璃块时都会发生弯曲。因为方块的两边是平行的，所以出射光平行于原来的入射光，但稍微偏移了一些。还要注意在玻璃块内表面的反射以及内部反射光线在上部表面的折射。

DISPERSION
色散

TERMS 术语

single wavelength	单波长
beam of white light	白光光束
visible wavelength	可见光波长
prism	棱镜
refraction of light by a prism	棱镜对光的折射
deviation of light by a prism	棱镜对光的偏折
prism spectroscope	棱镜分光镜
angle of deviation	偏折角
spectrum	光谱
series of colors	一系列颜色
violet light deviates the most	紫光偏折最大
red light deviates the least	红光偏折最小
index of refraction decreases with increasing wavelength	折射率随波长的增加而减小
apex angle of prism	棱镜顶角

LANGUAGE OF PHYSICS 物理学用语

Dispersion

The separation of white light into its component colors. It occurs because the index of refraction varies slightly with wavelength. In normal dispersion the refractive index of blue light is greater than that of red light. In anomalous dispersion it is the other way round.

色散
白光分解成不同颜色的光的现象。这是因为折射率随波长略有变化。在正常色散中，蓝光的折射率大于红光的折射率；在反常色散中，情况正好相反。

A **prism** is a transparent optical element having at least two polished plane faces inclined relative to each other, from which light is reflected or through which light is refracted.

棱镜：一种透明光学元件，它至少具有两个相对倾斜的抛光平面，光线从这些平面反射或穿过这些平面时发生折射。

The **angle of deviation** is the angle through which a ray of light is deviated by a refracting or reflecting surface, or a prism; the angle between an incident ray and the refracted or reflected ray.

偏折角：由于折射表面、反射表面或棱镜导致光线偏折所形成的角；入射光线与折射光线或反射光线之间的夹角。

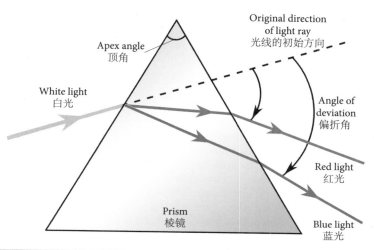

DISPERSION BY A PRISM

A prism refracts single-wavelength light and deflects the light through an angle of deviation. When a beam of white light (a combination of all visible wavelengths) is incident on a prism, the blue light is normally bent more than the red, although with certain materials this order can be reversed.

棱镜的色散

棱镜使单波长光发生折射，并使光偏转一个角度。当一束白光（所有可见波长的组合）入射到棱镜上时，蓝光通常比红光偏折更大，但在某些材料中，红光比蓝光偏折更大。

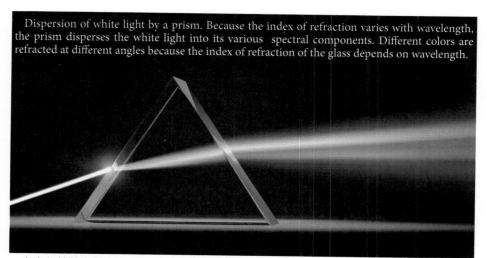

Dispersion of white light by a prism. Because the index of refraction varies with wavelength, the prism disperses the white light into its various spectral components. Different colors are refracted at different angles because the index of refraction of the glass depends on wavelength.

白光经棱镜色散。由于折射率随波长而变化，棱镜使白光色散，形成光谱。由于玻璃的折射率与波长相关，因此不同颜色的光折射角度不同。

HUYGENS' PRINCIPLE
惠更斯原理

TERMS	术语
propagation of light (wave)	光（波）的传播
wavefront	波前
reconstructed wavefront	重构波前
wavefront position	波前位置
point source	点源
primary point source or oscillator	一级点源或振子
secondary point source or oscillator	次级点源或振子
wave	波
plane wave	平面波
spherical wave	球面波
secondary wave	次级波
wavelet	子波
surface tangent to wavelet	和子波相切的表面
ray path	光路

LANGUAGE OF PHYSICS
物理学用语

Huygens' principle
All points on a given wavefront may be considered to be point sources (or secondary oscillators) for the production of spherical secondary waves, called wavelets, which propagate outward with speeds characteristic of waves in that medium. At any moment, the new position of the wavefront is the surface tangent to the wavelets.

惠更斯原理
一个已知的波前上的所有点都可以被看作产生次级球面波的点源（或次级振子），这种次级波被称为子波，以该介质中各处波的速度特性向外传播。在任何时候，波前的新位置就是和子波相切的表面。

A **wavefront** is an imaginary surface that represents corresponding points of a wave that vibrate in unison, i.e. with the same phase.

波前：一个假想的表面，代表了一列波中振动相同的点，即同相位的点。

In a **plane wave** the wavefront extends perpendicularly to the direction of propagation of the wavefront.

在**平面波**中，波前的延伸方向垂直于波前的传播方向。

In a **spherical wave** the wavefront travels radially outward in a direction away from the source.

对于**球面波**，波前从波源沿径向向外传播。

PLANE WAVE

Huygens' construction for a plane wave propagating to the right. Each point on the wavefront is considered as a point source (secondary oscillator) for generating wavelets.

平面波

向右传播的平面波的惠更斯构图。波前上的每个点都被看作产生子波的点源（次级振子）。

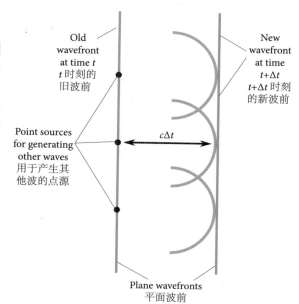

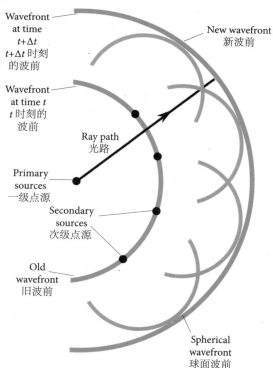

SPHERICAL WAVE

Huygens' construction for an outgoing spherical wave.

球面波

向外传播的球面波的惠更斯构图。

TOTAL INTERNAL REFLECTION
全反射

TERMS	术语
internal reflection	内反射
total internal reflection	全反射
multiple internal reflection	多重内反射
index of refraction	折射率
greater (high) index of refraction	高折射率
lower index of refraction	低折射率
boundary between media	介质分界面
interface	界面
incident angle	入射角
refracted angle	折射角
reflected angle	反射角
critical angle	临界角
light beam	光束
entirely reflected light beam	全反射光束
perfectly reflecting surface	最佳反射面
fiber optics	纤维光学
optical fiber	光纤
glass (plastic) rod	玻璃（塑料）棒
transparent rod	透明棒
telecommunication	电信

LANGUAGE OF PHYSICS
物理学用语

Total internal reflection occurs only when light attempts to move from a medium of a given index of refraction to a medium of lower index of refraction.

全反射只发生在光从高折射率介质向低折射率介质传播的过程中。

The **critical angle** is the angle of incidence that causes the refracted ray to have an angle of refraction of 90°. When the incident angle exceeds the critical angle no refraction occurs. In that case all the light that strikes the interface is reflected. It is said to undergo total internal reflection.

临界角：使折射光线的折射角为90°的入射角。入射角大于临界角时，不会发生折射。在那种情况下，所有照射在分界面上的光都被反射，即全反射。

Fiber optics
A flexible glass rod of high refractive index. Light entering the glass undergoes total internal reflection from the walls of the glass fiber and the light travels down the length of the fiber with little or no absorption of light.

纤维光学
光纤是一种高折射率的柔性玻璃棒。进入玻璃的光在玻璃纤维壁上发生全反射，沿着纤维的长度方向传播，几乎或完全不被吸收。

TOTAL INTERNAL REFLECTION

Rays from a medium 1 travel into a medium 2, where $n_2 < n_1$. As the angle of incidence increases, the refracted angle increases until it reaches 90° (ray 4). For this and larger angles of incidence, total internal reflection occurs (ray 5).

全反射

光线从介质1射入介质2，其中 n_2（光在介质2中的折射率）$< n_1$（光在介质1中的折射率）。随着入射角增大，折射角也增大，直到达到90°（射线4）。入射角大于或等于此角度时，发生全反射（射线5）。

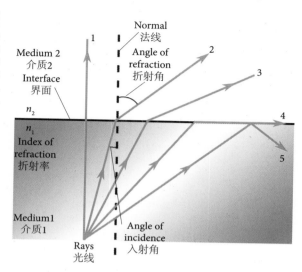

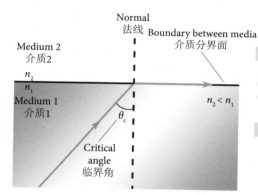

CRITICAL ANGLE

The angle of incidence producing an angle of refraction equals to 90° is the critical angle, θ_c.

临界角

使折射角等于90°的入射角就是临界角 θ_c。

FIBER OPTICS

Optical fibers serve multiple aplications, from telecommunication, to medical and industrial fields.

纤维光学

光纤被广泛应用于电信、医疗和工业领域。

IMAGES FORMED by MIRRORS
镜面成像

TERMS 术语

mirror	镜面
flat (plane) mirror	平面镜
spherical mirror	球面镜
concave mirror	凹面镜
convex mirror	凸面镜
radius (center) of curvature	曲率半径（中心）
image	像
real image	实像
virtual image	虚像
unmagnified image	等大的像
upright image	正像
diminished image	缩小的像
magnified image	放大的像
inverted image	倒像
image in front (back) of mirror	在镜面前（后）的像
principal axis	主轴
optical axis	光轴
point source of light	点光源
object distance	物距
viewer (observer)	观察者
image distance	像距
lateral magnification	横向放大率
paraxial ray	近轴光线
mirror equation	镜面方程
focal length	焦距
focal point	焦点
ray diagram	光路图

LANGUAGE OF PHYSICS
物理学用语

A **mirror** is a smooth, highly polished surface for reflecting light, that may be plane or curved. The actual reflecting surface is usually a thin coating of silver or aluminum on glass.

镜面：用来反射光线的光滑的、经过高度抛光的表面，它可以是平面的，也可以是曲面的。实际的反射面通常是玻璃上的一层薄薄的银或铝涂层。

A **spherical mirror** is a reflecting surface whose radius of curvature is the radius of the sphere from which the mirror is formed. A **plane mirror** is a special case of a spherical mirror with a radius of curvature that is infinite.

球面镜：一种反射面，其曲率半径等于形成镜面的球面半径。**平面镜**：球面镜的一种特殊情况，其曲率半径无穷大。

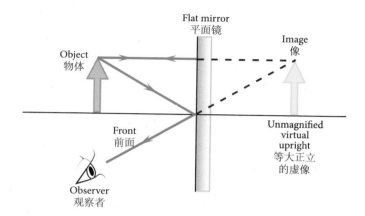

FLAT MIRROR
An image formed by reflection from a flat mirror. An object placed in front of a flat mirror. The image is located behind the mirror at a distance which is equal to the object distance.

平面镜
由平面镜反射而形成的像。平面镜前有一物体，其镜像位于镜面后方，像距等于物距。

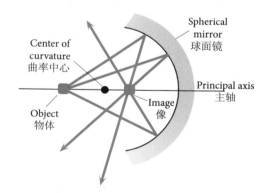

SPHERICAL MIRROR
A point object placed at a point on the principal axis in front of a concave spherical mirror forms a real image.

球面镜
位于凹面镜前主轴上的点状物体成实像。

A blind spot in a vehicle is an area around the vehicle that cannot be directly observed by the driver. A special spherical mirror reduces dangerous blind spots.

车辆上的盲点是指司机不能直接观察到的车辆周围的区域。图示为一种特殊球面镜，用以减少危险盲点。

THIN LENS
薄透镜

TERMS 术语

lens	透镜
converging lens	会聚透镜
diverging lens	发散透镜
lens thickest in the middle	中间最厚的镜片
lens thickest at the edge	边缘最厚的镜片
biconvex lens	双凸透镜
convex-concave lens	凸凹透镜
plano-convex lens	平凸透镜
biconcave lens	双凹透镜
plano-concave lens	平凹透镜
focal length	焦距
positive focal length	正焦距
negative focal length	负焦距
radius of curvature of lens surface	透镜表面的曲率半径
lens makers' equation	透镜成像公式
object	物体
image	像
object (image) focal point	物（像）方焦点
system of lens	透镜系统

LANGUAGE OF PHYSICS 物理学用语

A **lens** is a piece of transparent material, such as glass or plastic, that causes light passing through it to either converge or diverge using the process of refraction, depending on the shape of the material.

透镜：一片透明材料，如玻璃或塑料，通过改变材料形状，可使穿过的光线通过折射聚合或发散。

The **focal point** or **focus** is the point on the optical axis of a lens, to which an incident bundle of light rays parallel to the central axis of the lens will converge.

焦点：透镜光轴上一点，平行于透镜中心轴的入射光束会聚于此。

The **focal plane** is a plane through the focal point at right angles to the principal axis of a lens or mirror. The best image of distant objects is formed on that surface.

焦平面：通过焦点并与透镜或镜面主轴垂直的平面。远处物体在此平面上成像最佳。

The **principal axis** is a straight line connecting the curvature centers of the refracting lens surfaces. It is the optical axis of a lens.

主轴：一条连接折射透镜表面曲率中心的直线。主轴是透镜的光轴。

IMAGE FORMED BY A CONVERGING LENS

Ray diagram for locating the image formed by a thin lens. The object is located to the left of the object focal point of a converging lens.

会聚透镜成像

薄透镜成像光路图。物体位于会聚透镜物方焦点左侧。

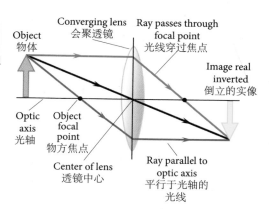

LENS SHAPES

(a) Converging lenses have a positive focal length and are thicker at the center than at the edge.

(b) Diverging lenses have a negative focal length and are thinner at the center than at the edge.

透镜形状

（a）会聚透镜焦距为正，中心比边缘厚。　　（b）发散透镜焦距为负，中心比边缘薄。

IMAGE FORMED BY A DIVERGING LENS

Ray diagram for locating the image formed by a thin lens. The object is located to the left of the object focal point of a diverging lens.

发散透镜成像

薄透镜成像光路图。物体位于发散透镜物方焦点左侧。

LENS ABERRATION
透镜像差

TERMS 术语

image	像
blurred image	模糊的像
imperfect image	非理想成像
sharp image	清晰的像
defects in shape of lenses	透镜形状缺陷
aberration	像差
spherical aberration	球差
chromatic aberration	色差
astigmatism	像散
coma	彗差
distortion	畸变
curvature of field	像场弯曲
center of lens	透镜中心
optical axis	光轴
focal plane	焦面
refractive index	折射率
wavelength of light	光的波长

LANGUAGE OF PHYSICS 物理学用语

Aberrations are defects of a lens system that cause the departure of real (imperfect) images from the ideal image predicted by simple ray theory. Aberrations cause the image formed by a lens to be blurred or distorted, with the nature of the distortion depending on the type of aberration. In an imaging system, it occurs when light from one point of an object does not converge to (or does not diverge from) a single point after transmission through the system.

像差：由于透镜系统缺陷，实际（非理想）成像与简单射线理论（几何光学）预测的理想成像之间存在的偏差。像差使透镜成像模糊或变形，如何变形取决于像差种类。在成像系统中，从物体上一点所发出的所有光线通过系统后未聚于一点时（或不从一点发散）时就会产生像差。

The **simple theory of mirrors and lenses** assumes that rays make small angles with the optical axis. All rays leaving a point source focus at a single point, producing a sharp image.

简单镜面和透镜理论：假设光线与光轴成小角度。离开点光源的所有光线聚焦于一点时，成像清晰。

SPHERICAL ABERRATION

Spherical aberration occurs when light passing through the lens at different distances from the optical axis is focused at different points.

球差

与光轴距离不同的光线通过透镜聚焦在不同的点上从而产生球差。

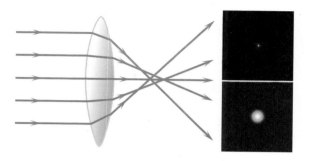

COMA

Coma occurs when light from an off-axis point source passing through the lens focuses at different points of the focal plane.

彗差

来自离轴点光源的光通过透镜在焦平面不同点聚焦而产生彗差。

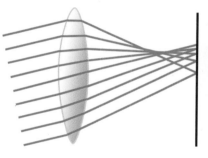 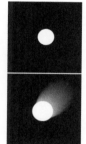

CHROMATIC ABERRATION

Chromatic aberration occurs because lenses have different refractive indices for different wavelengths of light.

色差

不同波长的光在透镜中折射率不同,从而产生色差。

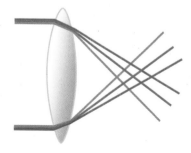 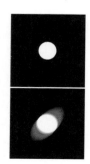

OPTICAL INSTRUMENT
光学仪器

TERMS	术语
eye	眼睛
- focuses light	-聚焦光
- produces a sharp image	-成像清晰
unaided (naked) eye	裸眼
cornea	角膜
aqueous humor	眼房水
iris	虹膜
pupil	瞳孔
crystalline lens	晶状体
retina	视网膜
rod cell	视杆细胞
cone cell	视锥细胞
accommodation	自动调焦
farsightedness (hyperopia)	远视
nearsightedness (myopia)	近视
power of a lens	透镜的焦强
simple magnifier	简单放大镜
magnification	放大率
angular magnification	角放大率
lateral magnification	横向放大率
confocal microscope	共焦显微镜
objective	物镜
eyepiece	目镜
telescope	望远镜
refracting telescope	折射式望远镜
reflecting telescope	反射式望远镜
parabolic mirror	抛物面镜
very distant object	非常遥远的物体
minute detail	微小的细节
spatial pinhole	空间针孔

LANGUAGE OF PHYSICS 物理学用语

An **optical image** is a reproduction of an object by an optical system. To describe an image it is important to specify its nature (real or virtual), its orientation (upright, or inverted) and its size (enlarged, true, or reduced).

光学成像：光学系统对物体的再现。要描述一个图像，重在指明其性质（实像或虚像）、方向（正立的或倒立的）和大小（放大的、原本的或缩小的）。

Magnification is the ratio of the size of an image to the size of its object.

放大率：像与物体的大小之比。

THE EYE

Essential parts of the eye. Like a camera, a normal eye focuses light and produces a sharp image.

眼睛

眼睛的主要部分如图所示。正常的眼睛像照相机一样聚焦光线，成像清晰。

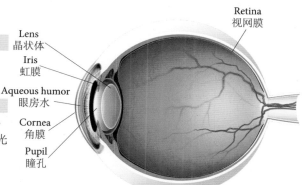

TELESCOPE

(a) The reflecting telescope uses a curved mirror and a lens. One of the largest reflectors in the world is the 9.8 m diameter Keck telescope in Hawaii.

(b) The refracting telescope uses a combination of lenses to form an image. The largest refracting telescope has a diameter of 1 m and is located at the Yerkes Observatory (USA).

望远镜

（a）反射式望远镜由曲面镜和透镜组成。世界上最大的反射镜之一是位于夏威夷的直径为9.8米的凯克望远镜。

（b）折射式望远镜使用多个透镜组合成像。最大的折射式望远镜直径为1米，位于耶基斯天文台（美国）。

CONFOCAL MICROSCOPY

Confocal microscopy is an optical imaging technique for increasing optical resolution and contrast of a micrograph by means of adding a spatial pinhole placed at the confocal plane of the lens to eliminate out-of-focus light.

共焦显微术

共焦显微术是一种光学成像技术，通过在透镜共焦平面上添加空间针孔来消除失焦光线，从而提高显微图像的光学分辨率和对比度。

INTERFERENCE of LIGHT WAVES
光波干涉

TERMS 术语

condition for interference	干涉条件
condition for constructive interference	相长干涉条件
condition for destructive interference	相消干涉条件
source of light	光源
coherent source of light	相干光源
incoherent source of light	非相干光源
monochromatic source of light	单色光源
path difference	光程差
superposition principle	叠加原理
constant phase relationship	恒定相位关系
interference pattern	干涉图样
fringe	条纹
bright fringe	明条纹
dark fringe	暗条纹
order number of fringe	条纹级数

LANGUAGE OF PHYSICS 物理学用语

Interference is a phenomenon in which two waves superpose to form a resultant wave of greater, lower, or equal amplitude.

干涉：两波叠加形成振幅更大、更小或相等的合成波的现象。

An **interference pattern** is observed if the sources are coherent and have identical wavelength, and the linear superposition principle is applicable.

干涉图样：如果光源相干且波长相同，则可以观察到干涉图样，且适用线性叠加原理。

The **optical path length** is the product of the geometrical distance travelled by light, and the index of refraction of the medium through which it propagates.

光程：光在介质中传播的几何距离与介质折射率之积。

The **path difference** is the difference in path travelled by the two waves measured in terms of the wavelength of the associated wave.

光程差：两列（相干）波走过的光程之差，由相应的波长来表示。

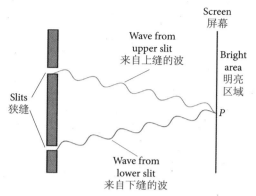

CONSTRUCTIVE INTERFERENCE

The two waves, which leave the two slits in phase, strike the screen at the central point P. Constructive interference occurs at this point and a bright area is observed.

相长干涉

两列波离开双缝时相位相同，在屏幕中心点 P 处干涉相长，形成亮条纹。

DESTRUCTIVE INTERFERENCE

Destructive interference occurs at point Q because the wave from the upper slit falls half a wavelength behind the wave from the lower slit and a dark region is observed at this point.

相消干涉

在 Q 点发生干涉相消，因为来自上缝的波比来自下缝的波在光程上落后半个波长，从而产生暗条纹。

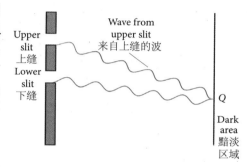

Interference effects observed in soap bubbles. The colors, produced just before the bubbles burst, are due to interference between light rays reflected from the front and back of the thin film of water making the bubble.

肥皂泡中的干涉效应。在气泡快要爆破时，光被薄水膜前后表面反射后发生干涉，从而呈现出不同颜色。

YOUNG'S DOUBLE-SLIT EXPERIMENT
杨氏双缝干涉实验

TERMS 术语

screen with two narrow parallel slits	有两条相互平行狭缝的屏幕
pair of two coherent light sources	相干光源对
intensity distribution of double-slit interference pattern	双缝干涉图样的强度分布
single-wavelength light source	单色光源
path difference	光程差

LANGUAGE OF PHYSICS 物理学用语

In **Young's double-slit experiment**, two slits separated by a fixed distance are illuminated by a single-wavelength light source. An interference pattern consisting of bright and dark fringes is observed on a viewing screen. The fringes can only be explained if light has the characteristics associated with waves.

杨氏双缝实验中，用单色光源照射两个间隔一定距离的狭缝。在显示屏上可以观察到由明暗条纹构成的干涉图样。只有光具有波动性质时，才能解释出现的条纹。

Fringes are a visible pattern of bright and dark parallel interference bands.

条纹：相互平行的明暗带状的可见干涉图样。

Monochromatic light consists of a single wavelength. In contrast, white light consists of light of many wavelengths.

单色光由单一波长组成。相反，白光由波长不同的光组成。

A **coherent light source** is a source of light that is capable of producing radiation with waves vibrating in phase.

相干光源：一种能够产生同相振动辐射波的光源。

The **laser** is an example of a coherent light source, as well as a source of monochromatic light.

激光器：一种相干光源，也是一种单色光源。

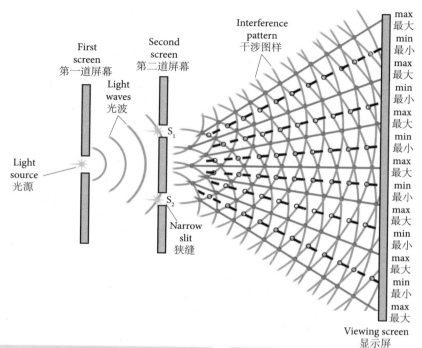

YOUNG'S DOUBLE-SLIT EXPERIMENT

Schematic diagram of Young's double-slit experiment. The narrow slits act as wave sources. Slits S_1 and S_2 behave as coherent sources that produce an interference pattern on the viewing screen.

杨氏双缝实验

杨氏双缝实验示意图。狭缝作为波源。狭缝S_1和S_2表现为相干源,在显示屏上产生干涉图样。

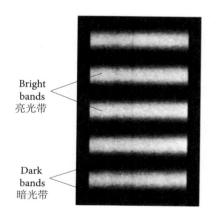

INTERFERENCE PATTERN

The light from the two slits produces on the viewing screen a visible pattern of bright and dark parallel bands called fringes.

干涉图样

两条狭缝发出的光在显示屏上产生相互平行、有明有暗、被称为条纹的带状可见图样。

INTERFERENCE in THIN FILM
薄膜干涉

TERMS	术语
thin film	薄膜
thin film of uniform thickness	等厚薄膜
index of refraction of thin film	薄膜折射率
difference in optical path length	光程差
phase change of 180°	180°相位差
no phase change	无相位差
phase change upon reflection	反射产生的相位差
path difference	（光）程差

LANGUAGE OF PHYSICS 物理学用语

A **thin film** is a very thin piece of transparent material. When monochromatic light strikes the film, interference effects can be observed due to reflection of light rays from the front and back of the thin film. The difference in the optical path of the two interfering light waves causes this effect.

薄膜：一种非常薄的透明材料。单色光照射在薄膜上，经由薄膜前后表面反射而发生干涉。这是由于两列相干波的光程不同造成的。

The **condition for constructive interference** in a film of thickness l and refractive index n with a common medium on both sides of the film (taking into account the phase change of 180° on reflection at one of the surfaces) is

在厚度为l、折射率为n的薄膜中，当薄膜两侧介质相同时（考虑到其中一个表面反射时产生180°的相位差），**相长干涉的条件**为

$$2nl = (m + \frac{1}{2})\lambda \quad (m = 0, 1, 2, ...)$$

The **condition for destructive interference** in a thin film is

薄膜中发生**相消干涉的条件**为

$$2nl = m\lambda \quad (m = 0, 1, 2, ...)$$

An **anti-reflective coating** means a thin-film coating, single or multilayer, that is applied to a substrate to decrease its reflectance over a specified range of wavelengths.

抗反射涂层：单层或多层的薄膜涂层应用于基材上以降低其在特定波长范围内的反射率。

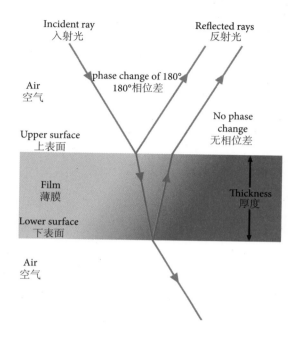

THIN FILM

Interference in light reflected from a thin film is due to a combination of rays reflected from the upper and lower surfaces of the film.

薄膜

薄膜反射光的干涉是由于经薄膜上下表面反射后光的叠加而产生的。

The brilliant colors in butterfly wings are due to interference. The structure of the wings causes constructive interference for a particular color such as blue. The color changes as you view the wings from different angles.

蝴蝶翅膀上绚丽的色彩是由干涉引起的。翅膀的结构会对特定的颜色（如蓝色）产生相长干涉。当你从不同角度观察翅膀时，颜色会发生变化。

NEWTON'S RINGS
牛顿环

TERMS 术语	
method for observing interference of light waves	光波干涉观测方法
plano-convex lens	平凸透镜
radius of curvature	曲率半径
flat glass surface	平板玻璃表面
point of contact	接触点
air film	空气薄膜
wavelength of light	光波长
circular fringe	圆条纹
bright circular fringe	亮圆条纹
dark circular fringe	暗圆条纹
radius of circular fringe	圆条纹半径
testing of optical lens	光学透镜测试

LANGUAGE OF PHYSICS 物理学用语

Newton's rings are a series of rings or bands formed when light beams reflected from two polished, adjacent surfaces, placed together with a thin film of air between them, interfere.

牛顿环：光线从两个相接触的光滑表面（两个表面之间有一层空气薄膜）反射而发生干涉形成的一系列环或带。

When **viewed with monochromatic light**, Newton's rings appear as a series of concentric, alternating bright and dark rings centered at the point of contact between the two surfaces.

用**单色光观察**时，牛顿环呈现出以两个（玻璃体）表面之间的接触点为中心的一系列明暗交替的同心圆环。

When **viewed with white light**, Newton's rings appear as a concentric ring pattern of rainbow colors, because the different wavelengths of light interfere at different thicknesses of the air layer between the surfaces. Measurements based on Newton's rings can be used to measure the refractive index of liquids between the surfaces.

用**白光观察**时，牛顿环呈现出彩色同心圆环图案，因为表面之间的空气层厚度不同，不同波长的光在不同厚度下发生干涉。牛顿环可以用来测量表面之间液体的折射率。

NEWTON'S RINGS

The interference effect is due to the combination of ray 1, reflected from the flat plate, with ray 2, reflected from the lower part of the lens. Ray 1 undergoes a phase change of 180° upon reflection, whereas ray 2 undergoes no phase change.

牛顿环

由平板玻璃反射出的光线1和由透镜下表面反射出的光线2相叠加发生干涉。光线1由于反射产生180°相位差,而光线2相位不变。

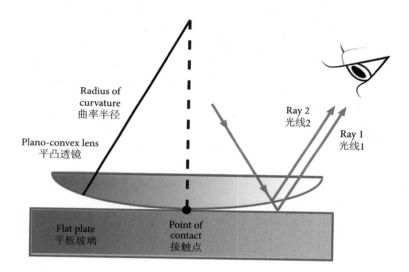

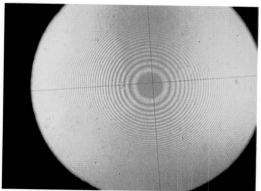

Photograph of Newton's rings. A circular pattern like this is obtained only when the lens is ground to a perfectly spherical curvature.

牛顿环照片。只有当透镜磨成完美的球面时,才能得到这样的圆形图案。

MICHELSON INTERFEROMETER
迈克尔逊干涉仪

TERMS	术语
light beam	光束
light beam split into two rays	光束分成两束
reflected ray	反射光
ray reflected vertically upward	垂直向上反射的光线
ray transmitted horizontally toward mirror	水平射向镜子的光线
two rays travel separate paths	两束光传播路径不同
two rays recombine to produce an interference pattern	两束光重新会聚产生干涉图样
mirror	镜子
adjustable mirror	动镜
half-silvered mirror	半涂银镜
LIGO Laser Interferometer Gravitational-Wave Observatory	LIGO激光干涉引力波天文台
precise measurement of displacement	位移精准测量

LANGUAGE OF PHYSICS   物理学用语

An **interferometer** is an instrument that splits a light beam into two parts and then recombines them to form an interference pattern.

干涉仪：将一束光分成两束再使其重新会聚形成干涉图样的仪器。

Application of interferometers
They can measure the accuracy of optical surfaces. These instruments employ the interference of light waves to measure length in terms of the length of a wave of light by using interference phenomena based on the wave characteristics of light. Interferometers are used for testing optical elements during manufacture.

干涉仪的应用
干涉仪可测量光学表面的精度。这些仪器基于光的波动特性，利用干涉现象，测量长度的精度可达光的波长级别。干涉仪可用于在制造过程中测试光学元件。

DIAGRAM OF MICHELSON INTERFEROMETER

A single light beam is split into two rays by the half-silvered mirror. The path difference between the two rays is varied with the adjustable mirror.

迈克尔逊干涉仪示意图

一束光被半涂银镜分成两束。两束光线之间的光程差可由动镜来调节。

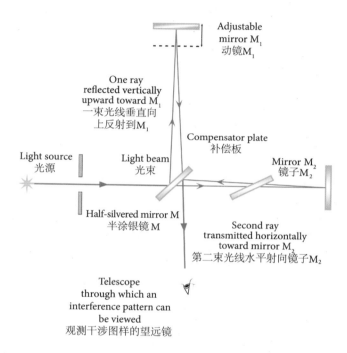

MICHELSON INTERFEROMETER

A basic Michelson interferometer. In 2015, another application of the Michelson interferometer, LIGO, made the first direct observation of gravitational waves.

迈克尔逊干涉仪

一台简易迈克尔逊干涉仪。2015年，迈克尔逊干涉仪的另一个应用——LIGO首次直接观测到引力波。

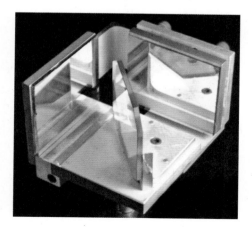

DIFFRACTION
衍射

TERMS	术语
diffraction	衍射
Fraunhofer diffraction	夫琅禾费衍射
Fresnel diffraction	菲涅耳衍射
single-slit diffraction	单缝衍射
divergence of light	光的发散
straight-line pass	直线传播
waves passing through small openings	穿过小孔的波
waves passing around obstacles	绕过障碍物的波
shadowed region	阴影区
diffraction pattern	衍射图样
diffraction angle	衍射角
intensity distribution	强度分布
condition for intensity minima	最小强度条件
Rayleigh's criterion	瑞利判据
angle resolution	角分辨率
circular aperture	圆孔

LANGUAGE OF PHYSICS
物理学用语

Diffraction is the bending of light around an obstacle, into a region that should be in shadow.

衍射：光绕过障碍物进入原本应该是阴影区域的现象。

A **diffraction pattern** consists of a broad, intense central band called the central maximum, flanked by a series of narrower, less intense additional bands, called side maxima or secondary maxima and a series of intervening dark bands (or minima).

衍射图样：该图样由中央主极大、两边的一系列次（边）极大和一系列极弱（暗）条纹构成。其中，中央条纹最宽也最亮，两侧明条纹相对较窄也较暗，且明条纹间穿插有暗条纹。

The **diffraction angle** is the angle between the direction of an incident light beam and any resulting diffracted beam.

衍射角：入射光束和任意一束衍射光线之间的夹角。

The **Rayleigh's criterion** for the diffraction limit to resolution states that two images are just resolvable when the center of the diffraction pattern of one is directly over the first minimum of the diffraction pattern of the other.

衍射极限分辨率的**瑞利判据**：一个衍射图样的中心恰好与另一个衍射图样的一级暗纹重合时，刚好能分辨出是两个像。

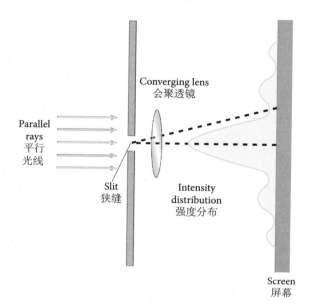

FRAUNHOFER DIFFRACTION

Fraunhofer diffraction occurs when the rays reaching a viewing screen are approximately parallel. The screen is far from the slit. A converging lens is used to focus the parallel rays on the screen.

夫琅禾费衍射

到达显示屏的光线近似平行时，就会发生夫琅禾费衍射。屏幕离狭缝很远，会聚透镜用于将平行光线聚焦到屏幕上。

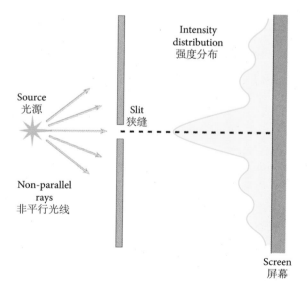

FRESNEL DIFFRACTION

A Fresnel diffraction pattern of a single slit is observed when the rays are not parallel and the observing screen is a finite distance from the slit.

菲涅尔衍射

光线不平行且显示屏和单缝之间距离有限时，可以观察到单缝菲涅尔衍射图样。

DIFFRACTION GRATING
衍射光栅

TERMS 术语

grating	光栅
diffraction grating	衍射光栅
transmission grating	透射光栅
reflection grating	反射光栅
slit spacing	狭缝间距
slit	狭缝
parallel slits	平行狭缝
equally spaced slits	等距狭缝
transparent slits	透明狭缝
closely spaced slits	密狭缝
diffracted beam	衍射光束
diffracted beams interfere with each other	衍射光束互相干涉
diffracted beams produce final pattern	衍射光束形成最终图样
diffracted beams as a result of interference and diffraction	干涉和衍射产生的衍射光束
path difference	光程差
angle of deviation	偏折角
maximum of mth order	第 m 级极大
principal maximum	主极大
resolving power	分辨能力
ruling (groove)	刻线（凹槽）

LANGUAGE OF PHYSICS   物理学用语

A **diffraction grating** is a device consisting of a large number of equally spaced parallel slits that gives a characteristic intensity distribution of sharper and narrower fringes.

衍射光栅：一种由大量的等距平行狭缝组成的装置。其强度分布的特点是条纹更加清晰、更加狭窄。

When a plane wave is incident normal to the plane of the grating, the **diffraction pattern** observed on the screen is the result of the combined effects of interference and diffraction. Each slit produces diffraction, and the diffracted beams interfere with one another to produce the final pattern.

平面波垂直于光栅平面入射时，在屏幕上观察到的**衍射图样**是干涉和衍射共同作用的结果。每一个狭缝都会产生衍射，衍射光束又会互相干涉，从而形成最终的图样。

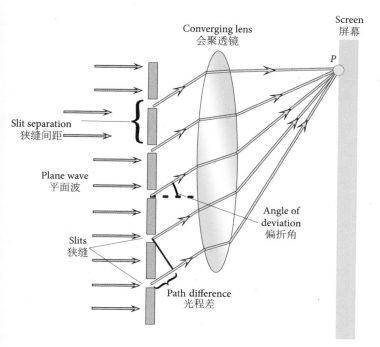

DIFFRACTION GRATING

Side view of a diffraction grating. If the path difference equals one wavelength or some integer multiple of a wavelength, waves from all slits are in phase at P and a bright line is observed.

衍射光栅

衍射光栅的侧视图。光程差等于一个波长或波长的整数倍时,来自所有狭缝的波在点P处同相位,由此就能观察到一条亮线。

A reflective diffraction grating is a periodic structure patterned into a reflective surface, which disperses light into its constituent wavelengths (separates light into its colors). Used in all applications in which the wavelength of light must be tuned (spectroscopy, imaging, etc.), including beamline monochromators and X-ray emission spectrometers.

反射衍射光栅是一种具有周期性结构的反射面,可以用来将一束光按其组分波长进行分散(即将光分解成单色光)。衍射光栅应用于需要调整波长的仪器(如在光谱学、成像技术中的应用等),包括光束线单色仪和X射线发射光谱仪。

DIFFRACTION of X-RAY by CRYSTAL
晶体的X射线衍射

TERMS	术语
X-ray	X射线
X-ray beam	X射线束
electromagnetic wave of short wavelength	短波长电磁波
atomic spacing	原子间距
array of atoms	原子阵列
three-dimensional diffraction grating	三维衍射光栅
crystalline structure	晶体结构
crystalline plane	晶面
discrete crystalline plane	离散晶面
upper crystalline plane	上层晶面
lower crystalline plane	下层晶面
Laue pattern	劳厄图谱
Bragg's law	布拉格定律

LANGUAGE OF PHYSICS   物理学用语

X-ray diffraction means the bending of X-rays by regular layers of molecules in a crystal acting like a very small diffraction grating. The diffraction pattern so obtained and recorded on film provides a means for analyzing the crystal structure.

X射线衍射：晶体中规则排列的分子层就像非常小的衍射光栅，X射线穿过晶体会发生弯曲。因此在底片上形成并记录下来的衍射图样能够用来分析晶体结构。

Bragg's law expresses the condition under which a crystal will reflect a beam of X-rays with the greatest amount of resolution. This law denotes the **Bragg angle** formed by the incident beam and the crystal planes, at which the reflection occurs.

布拉格定律：给出了X射线束由晶体反射后获得最佳分辨率时所需满足的条件。该定律还说明了入射光束和发生反射的晶面之间的**布拉格角**。

X-ray crystallography is a technique used for determining the atomic and molecular structure of a crystal, in which the crystalline structure causes a beam of incident X-rays to diffract into many specific directions. It can be used for determining the mean positions of the atoms in the crystal, their chemical bonds and other types of information.

X射线晶体学：一种用于确定晶体中原子和分子结构的技术。入射的X射线束会被晶体结构衍射到一些特定的方向上，从而可以确定晶体中原子的平均位置、化学键等信息。

LAUE PATTERN

Schematic diagram of the technique used to observe the diffraction of X-rays by a crystal. The array of spots formed on the film is called a Laue pattern.

劳厄图谱

晶体X射线衍射技术示意图。底片上形成的斑点阵列称为劳厄图谱。

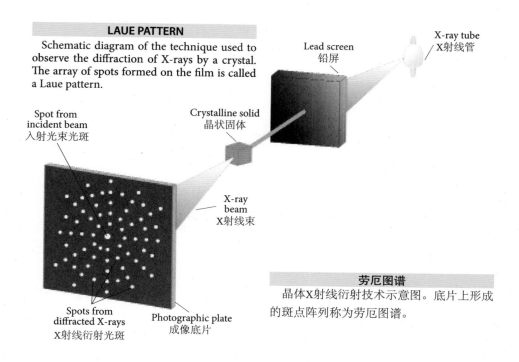

BRAGG'S LAW

A two-dimensional description of the reflection of an X-ray beam from two parallel crystalline planes.

布拉格定律

X射线束被两个平行晶面反射的二维示意图。

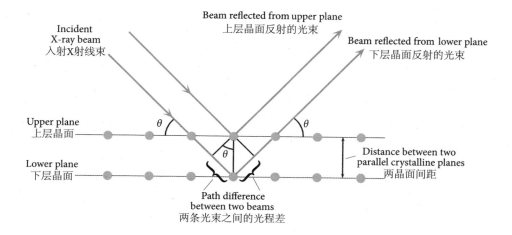

POLARIZATION of LIGHT WAVE
光的偏振

TERMS 术语

light wave	光波
unpolarized (natural) light wave	非偏振（自然）光波
polarized light wave	偏振光波
plane-polarized light wave	面偏振光波
linearly polarized light wave	线偏振光波
elliptically polarized light wave	椭圆偏振光波
direction of polarization	偏振方向
direction of wave propagation	波的传播方向
direction of electric field vector	电场矢量方向
plane of polarization	偏振面
polarization by selective absorption	选择性吸收起偏
polarization by reflection	反射起偏
polarization by double refraction	双折射起偏
polarization by scattering	散射起偏
polaroid	偏振片
degree of polarization	偏振度

LANGUAGE OF PHYSICS   物理学用语

Polarization is the property of a wave that can oscillate with more than one orientation. A light wave that is vibrating in more than one plane is referred to as **unpolarized light**. The process of transforming unpolarized light into polarized light is known as **polarization of light**.

偏振：波能够在多个方向上振动的一种特性。在不同平面上振动的光波称为**非偏振光**。非偏振光转化为偏振光的过程称为**光的极化（偏振）**。

A light wave is said to be **linearly polarized** if the electric field vector vibrates in a single plane perpendicular to the direction of propagation.

如果一束光的电场矢量在垂直于传播方向的平面内振动，那么就称这束光为**线偏振光**。

Plane of polarization is the plane formed by electric field vector E and the direction of propagation.

偏振面：电场矢量E以及光的传播方向形成的平面。

The **direction of polarization** of each individual wave is defined to be the direction in which the electric field is vibrating.

光波的**偏振方向**：电场的振动方向。

UNPOLARIZED LIGHT

An unpolarized light beam viewed along the direction of propagation (perpendicular to the page). The transverse electric field vector can vibrate in any direction with equal probability.

POLARIZED LIGHT

A linearly polarized light beam with the electric field vector vibrating in the vertical direction.

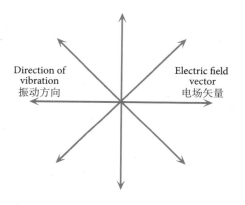

非偏振光

沿传播方向（垂直纸面的方向）观察非偏振光。横向电场矢量在任意方向上等概率振动。

偏振光

电场矢量沿垂直方向振动的线偏振光。

The effects of a polarizing filter (left image). Unpolarized light, after reflection at a specular (shiny) surface, generally obtains a degree of polarization. Polarizing sunglasses exploit this effect to reduce glare from reflections by horizontal surfaces, notably the road ahead viewed at a grazing angle.

偏振滤光片的效果（左图）。非偏振光被镜面（或光亮的表面）反射后，通常会产生一定程度的偏振。偏光太阳镜就是利用这一原理来减少地面反射的眩光，尤其是以掠射角看向前方路面时产生的眩光。

MALUS' LAW
马吕斯定律

TERMS	术语
technique for polarizing light	偏光技术
polarizer	起偏器
analyzer	检偏器
transmission (polarizer) axis	透射（偏振）轴
intensity of transmitted light	透射光的强度
intensity of polarized light	偏振光的强度
intensity of polarized wave	偏振波的强度
wave incident on analyzer	入射在检偏器上的波

LANGUAGE OF PHYSICS 物理学用语

A **polarizer** is an optical device capable of transforming unpolarized or natural light into polarized light.

起偏器：一种能够将非偏振光（自然光）转换为偏振光的光学装置。

An **analyzer** is an optical device capable of producing plane-polarized light, and used for detecting the effect of the object on plane-polarized light produced by the polarizer.

检偏器：一种能够产生面偏振光的光学装置，它用于检测物体对起偏器产生的平面偏振光的影响。

Polaroid is a material that polarizes light.

偏振片：一种使光发生偏振的材料。

Malus' law
When polarized light of intensity I_0 is incident on a polarizer, the light transmitted through the polarizer has an amplitude $E_0 \cos\theta$ and an intensity equal to $I_0 \cos^2\theta$, where θ is the angle between the transmission axis of the polarizer and the electric field vector of the incident light.

马吕斯定律
强度为 I_0 的偏振光穿过起偏器后，其振幅会变为 $E_0\cos\theta$，强度为 $I_0\cos^2\theta$，其中 θ 为起偏器透射轴与入射光电场矢量之间的夹角。

MALUS' LAW

Two polarizing sheets whose transmission axes make an angle θ with each other. Only a fraction of the polarized light incident on the analyzer is transmitted.

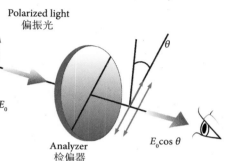

马吕斯定律

两个偏振片的透射轴之间的夹角为θ。只有部分偏振光能透过检偏器。

LIGHT TRANSMITTED THROUGH POLARIZERS

The intensity of light transmitted through two polarizers depends on the relative orientation of their transmission axes.

穿过起偏器的光

穿过两个起偏器后，光的强度取决于两个起偏器透射轴的相对方位。

(a) The transmitted intensity is maximum when the transmission axes are parallel ($\theta = 0$ or $180°$).

（a）透射轴互相平行（即$\theta=0$或者$180°$）时，透射强度最大。

(b) The transmitted light intensity diminishes when the transmission axes are at an angle of $45°$ with each other.

（b）透射轴之间夹角为$45°$时，透射强度减小。

(c) The transmitted intensity is zero when the transmission axes are perpendicular to each other.

（c）透射轴互相垂直时，透射强度为0。

BIREFRINGENCE
双折射

TERMS 术语	
crystalline material	晶体材料
amorphous material	非晶材料
double-refracting material	双折射材料
birefringent material	双折射材料
anisotropic material	各向异性材料
single index of refraction	单折射率
two indices of refraction	双折射率
ordinary ray	寻常光
extraordinary ray	非常光
optic axis	光轴
calcite	方解石

LANGUAGE OF PHYSICS 物理学用语

Double refraction, also called **birefringence**, is an optical property in which a single ray of unpolarized light entering an anisotropic medium is split into two plane-polarized rays. One ray (called the extraordinary ray) travels with different speeds in different directions, the other ray (called the ordinary ray) has the same speed in all directions and passes through the medium unchanged.

A **Nicol prism** is an analyzer. It made of calcite (Iceland spar) that has been cut at an angle of 68° with respect to the crystal axis, cut again diagonally, and then rejoined using a layer of transparent Canada balsam as a glue. Unpolarized light striking an end face is polarized to vibrate in two directions. The extraordinary ray passes through the crystal, and the ordinary ray is reflected to one side at the calcite-balsam interface and is lost.

双折射：一种光学特性，即非偏振光的单束光进入各向异性介质中被分离成两束面偏振光的现象。其中一束光（称为非常光，简称e光）在不同方向上的传播速度不同，而另一束光（称为寻常光，简称o光）在所有方向上的传播速度都相同，且在介质中传播时保持不变。

尼科尔棱镜：一种检偏器，原材料是方解石（冰岛晶石）。制作时先沿与晶轴成68°角的方向切割，再沿对角线切割，然后用一层透明的加拿大树胶作为胶水将两片晶体黏合起来。非偏振光从棱镜一端入射后会发生偏振而在两个方向振动。非常光会穿过方解石，而寻常光则在晶体-树胶交界处被反射，随后消失。

BIREFRINGENCE

Unpolarized light incident on a calcite crystal splits into an ordinary ray and an extraordinary ray. These two rays are polarized in mutually perpendicular directions.

双折射

入射到方解石晶体上的非偏振光分离为寻常光和非常光。这两束光在相互垂直的方向上发生偏振。

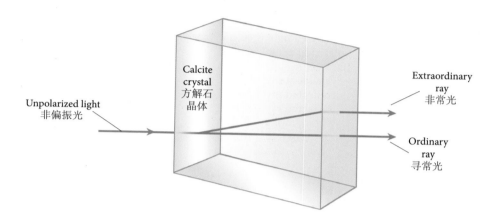

A calcite crystal laid upon a paper with black line and word "Calcite" showing double refraction.

在一张纸上画上黑色的线条并写上"Calcite",再将方解石放在纸上即可观察到双折射现象。

BREWSTER'S LAW
布儒斯特定律

TERMS 术语

light	光
reflected light	反射光
polarized light	偏振光
completely polarized light	完全偏振光
partially polarized light	部分偏振光
unpolarized light	非偏振光
polarizing angle	起偏角
Brewster's angle	布儒斯特角
incident ray	入射光
reflected ray	反射光
refracted ray	折射光
holography	全息术
hologram	全息影像

LANGUAGE OF PHYSICS 物理学用语

Brewster's law
When light strikes a surface at such an angle that the reflected and refracted rays are perpendicular to each other, the reflected light is completely polarized.

The angle of incidence, called the **polarizing angle** θ, satisfies the condition $n = \tan \theta$, where n is the index of refraction of the reflecting medium.

布儒斯特定律
当一束光照射在表面上产生的反射光和折射光互相垂直时，其中的反射光将发生完全偏振。

此时的入射角被称为**起偏角**θ，满足$n=\tan\theta$，其中n为反射介质的折射率。

BREWSTER'S LAW

The reflected beam is completely polarized when the incident angle equals the polarizing angle, which satisfies Brewster's law.

布儒斯特定律

当入射角等于起偏角,即满足布儒斯特定律时,反射光束发生完全偏振。

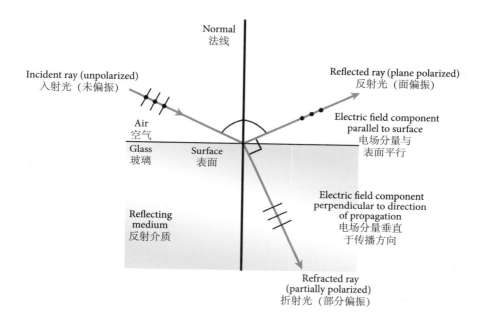

The hologram of the Golden Pectoral. This is an ancient Scythian treasure discovered in a burial kurgan in Ukraine. When recording a hologram, light is typically incident at Brewster's angle. Holography is a photographic technique that records the light scattered from an object, and then presents it in a way that appears three-dimensional.

黄金胸饰的全息影像。在乌克兰一个墓葬中,人们发现了这一古老的西西亚宝藏。利用全息图记录它时,光通常以布儒斯特角入射。全息技术是一种摄影技术,这种技术能够记录物体散射出来的光,然后将物体以三维的方式呈现出来。

APPENDICES
附录

GREEK ALPHABET and its COMMON USAGE — 294
希腊字母表及常见用法

CHEMICAL SYMBOLS for the ELEMENTS — 296
化学元素符号

SELECTED FUNDAMENTAL CONSTANTS — 300
常用基本常数

PHYSICAL QUANTITIES — 302
物理量

SI BASE UNITS and DERIVED UNITS — 306
国际单位制基本单位和导出单位

SELECTED PREFIXES for POWERS of TEN — 307
常用10的幂次前缀

MATHEMATICAL SYMBOLS and OPERATIONS — 308
数学符号和运算

ALGEBRA — 309
代数

GEOMETRY — 310
几何

TRIGONOMETRY — 311
三角函数

GREEK ALPHABET and its COMMON USAGE

希腊字母表及常见用法

Capital (cap)/ 大写	Lower case (lc)/ 小写	Name	名称	Commonly Used to Designate	常用于表示
A	α	alpha	阿尔法	Angle, angular acceleration, coefficient(lc)	角度（小写）、角加速度（小写）、系数（小写）
B	β	beta	贝塔	Angle, coefficient(lc)	角度（小写）、系数（小写）
Γ	γ	gamma	伽马	Specific gravity, angle, electrical conductivity, surface tension (lc)	比重（小写）、角度（小写）、电导率（小写）、表面张力（小写）
Δ	δ	delta	德尔塔	Change in something (cap), decrement (lc), determinant (cap), angle (lc)	变化量（大写）、减小量（小写）、行列式（大写）、角度（小写）
E	ε	epsilon	艾普西隆	Dielectric constant, permittivity of free space, emissivity (lc)	介电常数（小写）、真空介电常数（小写）、辐射率（小写）
Z	ζ	zeta	泽塔	Coordinate, coefficients (lc)	坐标、系数（小写）
H	η	eta	伊塔	Coefficient of viscosity, efficiency (lc)	黏滞系数（小写）、功率（小写）
Θ	θ	theta	西塔	Angle, angular displacement (lc)	角度（小写）、角位移（小写）
I	ι	iota	约塔	Unit vector (lc)	单位矢量（小写）
K	κ	kappa	卡帕	Thermal conductivity (lc)	热导率（小写）
Λ	λ	lambda	拉姆达	Wavelength, linear charge density (lc)	波长（小写）、线电荷密度（小写）

GREEK APLHABET and its COMMON USAGE

希腊字母表及常见用法

Capital (cap)/ 大写	Lower case (lc)/ 小写	Name	名称	Commonly Used to Designate	常用于表示
Μ	μ	mu	谬	Coefficient of friction, permeability, prefix micro, molar mass (lc)	摩擦系数（小写）、磁导率（小写）、前缀"微"（小写）、摩尔质量（小写）
Ν	ν	nu	纽	Frequency, number of moles (lc)	频率（小写）、摩尔数（小写）
Ξ	ξ	xi	克西	Coordinate (lc)	坐标（小写）
Ο	ο	omicron	奥米克戎	—	—
Π	π	pi	派	3.141 6	3.141 6
Ρ	ρ	rho	柔	Density (volume charge density), resistivity (lc)	密度（体电荷密度）（小写）、电阻率（小写）
Σ	σ	sigma	西格马	Summation (cap), surface charge density, electrical conductivity (lc)	求和（大写）、面电荷密度（小写）、电导率（小写）
Τ	τ	tau	陶	Torque, time constant (lc)	力矩（小写）、时间常数（小写）
Υ	υ	upsilon	宇普西隆	—	—
Φ	φ	phi	斐	Potential, angle (lc), magnetic flux (cap)	电势、角度（小写），磁通量（大写）
Χ	χ	chi	希	Electric susceptibility (lc)	电极化率（小写）
Ψ	ψ	psi	普西	Wave function (lc)	波函数（小写）
Ω	ω	omega	奥米伽	Solid angle (cap), resistance (cap), angular velocity, angular frequency	立体角（大写）、阻力（大写）、角速度、角频率

CHEMICAL SYMBOLS for the ELEMENTS

化学元素符号

Symbol/ 符号	Name	名称	Atomic No./ 原子序数
Ac	Actinium	锕	89
Ag	Silver	银	47
Al	Aluminum	铝	13
Am	Americium	镅	95
Ar	Argon	氩	18
As	Arsenic	砷	33
At	Astatine	砹	85
Au	Gold	金	79
B	Boron	硼	5
Ba	Barium	钡	56
Be	Beryllium	铍	4
Bi	Bismuth	铋	83
Bk	Berkelium	锫	97
Br	Bromine	溴	35
C	Carbon	碳	6
Ca	Calcium	钙	20
Cd	Cadmium	镉	48
Ce	Cerium	铈	58
Cf	Californium	锎	98
Cl	Chlorine	氯	17
Cm	Curium	锔	96
Co	Cobalt	钴	27
Cr	Chromium	铬	24
Cs	Cesium	铯	55
Cu	Copper	铜	29
Dy	Dysprosium	镝	66
Er	Erbium	铒	68

CHEMICAL SYMBOLS for the ELEMENTS

化学元素符号

Symbol/ 符号	Name	名称	Atomic No./ 原子序数
Es	Einsteinium	锿	99
Eu	Europium	铕	63
F	Fluorine	氟	9
Fe	Iron	铁	26
Fm	Fermium	镄	100
Fr	Francium	钫	87
Ga	Gallium	镓	31
Gd	Gadolinium	钆	64
Ge	Germanium	锗	32
H	Hydrogen	氢	1
He	Helium	氦	2
Hf	Hafnium	铪	72
Hg	Mercury	汞	80
Ho	Holmium	钬	67
I	Iodine	碘	53
In	Indium	铟	49
Ir	Iridium	铱	77
K	Potassium	钾	19
Kr	Krypton	氪	36
La	Lanthanum	镧	57
Li	Lithium	锂	3
Lr	Lawrencium	铹	103
Lu	Lutetium	镥	71
Md	Mendelevium	钔	101
Mg	Magnesium	镁	12
Mn	Manganese	锰	25
Mo	Molybdenum	钼	42

CHEMICAL SYMBOLS for the ELEMENTS

化学元素符号

Symbol/符号	Name	名称	Atomic No./原子序数
N	Nitrogen	氮	7
Na	Sodium	钠	11
Nb	Niobium	铌	41
Nd	Neodymium	钕	60
Ne	Neon	氖	10
Ni	Nickel	镍	28
No	Nobelium	锘	102
Np	Neptunium	镎	93
O	Oxygen	氧	8
Os	Osmium	锇	76
P	Phosphorus	磷	15
Pa	Protactinium	镤	91
Pb	Lead	铅	82
Pd	Palladium	钯	46
Pm	Promethium	钷	61
Po	Polonium	钋	84
Pr	Praseodymium	镨	59
Pt	Platinum	铂	78
Pu	Plutonium	钚	94
Ra	Radium	镭	88
Rb	Rubidium	铷	37
Re	Rhenium	铼	75
Rh	Rhodium	铑	45
Rn	Radon	氡	86
Ru	Ruthenium	钌	44
S	Sulfur	硫	16
Sb	Antimony	锑	51

CHEMICAL SYMBOLS for the ELEMENTS

化学元素符号

Symbol/符号	Name	名称	Atomic No./原子序数
Sc	Scandium	钪	21
Se	Selenium	硒	34
Si	Silicon	硅	14
Sm	Samarium	钐	62
Sn	Tin	锡	50
Sr	Strontium	锶	38
Ta	Tantalum	钽	73
Tb	Terbium	铽	65
Tc	Technetium	锝	43
Te	Tellurium	碲	52
Th	Thorium	钍	90
Ti	Titanium	钛	22
Tl	Thallium	铊	81
Tm	Thulium	铥	69
U	Uranium	铀	92
V	Vanadium	钒	23
W	Tungsten	钨	74
Xe	Xenon	氙	54
Y	Yttrium	钇	39
Yb	Ytterbium	镱	70
Zn	Zinc	锌	30
Zr	Zirconium	锆	40

SELECTED FUNDAMENTAL CONSTANTS

常用基本常数

Quantity	量	Symbol/符号	Value/数值	Unit	单位
Atomic mass unit	原子质量单位	u	1.66054×10^{-27}	kg	千克
Avogadro's number	阿伏伽德罗常数	N_A	$6.02214076 \times 10^{23}$	mol^{-1}	摩尔$^{-1}$
Bohr magneton	玻尔磁子	μ_B	9.27401×10^{-24}	J/T	焦耳/特斯拉
Bohr radius	玻尔半径	a_0	5.2918×10^{-11}	m	米
Boltzmann's constant	玻尔兹曼常数	k_B	1.38066×10^{-23}	J/K	焦耳/开尔文
Compton wavelength	康普顿波长	λ_C	2.42631×10^{-12}	m	米
Deuteron mass	氘核质量	m_d	3.34359×10^{-27}	kg	千克
Electron mass	电子质量	m_e	9.10939×10^{-31}	kg	千克
Electron-volt	电子伏	eV	1.60218×10^{-19}	J	焦耳
Elementary charge	元电荷	e	1.60218×10^{-19}	C	库仑
Faraday constant	法拉第常数	F	96 520	C/mol	库仑/摩尔
Gas constant	气体常数	R	8.314 511	J/(mol·K)	焦耳/(摩尔·开尔文)
Gravitational constant	引力常数	G	6.67260×10^{-11}	$N \cdot m^2/kg^2$	牛顿·米^2/千克^2
Hydrogen ground state energy	氢原子基态能量	E_0	13.605 698	eV	电子伏
Josephson frequency-voltage ratio	约瑟夫森频率-电压比	$2e/h$	4.835 977	Hz/V	赫兹/伏特

SELECTED FUNDAMENTAL CONSTANTS

常用基本常数

Quantity	量	Symbol/符号	Value/数值	Unit	单位
Magnetic flux quantum	磁通量子	Φ_0	$2.067\,835 \times 10^{-15}$	Wb	韦伯
Neutron mass	中子质量	m_n	$1.674\,929 \times 10^{-27}$	kg	千克
Nuclear magneton	核磁子	μ_n	$5.050\,787 \times 10^{-27}$	J/T	焦耳/特斯拉
Permeability of free space	真空磁导率	μ_0	$4\pi \times 10^{-7}$	N/A^2	牛顿/安培^2
Permittivity of free space	真空介电系数	ε_0	$8.854\,188 \times 10^{-12}$	$C^2/(N \cdot m^2)$	库仑^2/(牛顿·米^2)
Planck's constant	普朗克常数	h	$6.626\,070\,15 \times 10^{-34}$	J·s	焦耳·秒
Proton mass	质子质量	m_p	$1.672\,623 \times 10^{-27}$	kg	千克
Quantized Hall resistance	量子化霍尔电阻	h/e^2	$25\,812.806$	Ω	欧姆
Rydberg constant	里德伯常量	R_H	$1.097\,373 \times 10^{7}$	m^{-1}	米$^{-1}$
Specific elementary charge	比元电荷	e/m	$1.758\,88 \times 10^{11}$	C/m	库仑/米
Speed of light in a vacuum	真空中的光速	c	$2.997\,925 \times 10^{8}$	m/s	米/秒
Standard acceleration due to gravity (standard gravitational field strength)	标准重力加速度（标准重力场强度）	g	$9.806\,65$	m/s^2 (N/kg)	米/秒^2 (牛顿/千克)
STP= standard temperature and pressure	标准温度和标准压强		0 760	°C mmHg	摄氏度 毫米汞柱

PHYSICAL QUANTITIES

物理量

Definition	定义
The **meter**, **m**, is the SI base unit of length. The meter is defined as the length of the path travelled by light in a vacuum in 1/299 792 458 of a second.	米（m）：国际单位制中长度的基本单位，是真空中光在1/299 792 458秒内传播的路径长度。
The **kilogram**, **kg**, is the SI base unit of mass. It is defined by fixing the Planck's constant, h, to be exactly $6.626\ 070\ 15 \times 10^{-34}$ J·s, which is linked to kg via the base units as kg·m²·s⁻¹, and meters and seconds are the SI base units for distance and time.	千克（kg）：国际单位制中质量的基本单位。取普朗克常数h为一定值——$6.626\ 070\ 15 \times 10^{-34}$焦耳·秒，再利用该常数将千克与其他基本单位通过千克·米²/秒联系起来，其中米和秒为国际单位制中距离和时间的基本单位。
The **second**, **s**, is the SI base unit of time. One second is defined as the duration of 9 192 631 770 periods of the radiation corresponding to the transition between the two hyperfine levels of the unperturbed ground state of the cesium-133 atom.	秒（s）：国际单位制中时间的基本单位，将1秒定义为铯-133原子两个超精细能级之间跃迁对应辐射的9 192 631 770个周期所持续的时间。
The **Ampere**, **A**, is the SI base unit of electric current. It is defined by fixing the elementary electrical charge, e, as $1.602\ 176\ 634 \times 10^{-19}$ coulomb (C), which is equal to A·s, and s is the SI base unit of time.	安培（A）：国际单位制中电流的基本单位。取基本电荷e为一定值——$1.602\ 176\ 634 \times 10^{-19}$库仑（C），库仑相当于A·s（安培·秒），其中秒（s）是国际单位制中时间的基本单位。
The **Kelvin**, **K**, is the SI base unit of temperature and is 1/273.16 of the temperature difference between 0 K and the triple point of water, at which ice, water and water vapour coexist. The melting point of pure ice is 273.15 K.	开尔文（K）：国际单位制中温度的基本单位，是0 K和水三相点之间差值的1/273.16。其中水三相点是冰、水、水蒸气可以共存的温度。纯冰的熔点是273.15 K。

PHYSICAL QUANTITIES

物理量

Definition	定义
The **mole, mol**, is the SI base unit for the amount of substance. One mole contains exactly 6.022 140 76 × 10^{23} elementary entities. This number defines the numerical value of the Avogadro constant, N_A, when expressed in the unit mol^{-1} and it is called the Avogadro's number. An elementary entity may be an atom, a molecule, an ion, an electron, or any other particle or specified group of particles.	摩尔（mol）：国际单位制中物质的量的基本单位。1摩尔恰好包含6.022 140 76×10^{23}个基本单元。这一数字定义了阿伏伽德罗常量 N_A 的数值，单位为摩尔$^{-1}$，该数被称为阿伏伽德罗常数。基本单元可以是原子、分子、离子、电子或任何其他粒子或者特定的粒子团。
The **candela, cd**, is the SI base unit for luminous intensity and is defined as an intensity of 1/683 watt·steradian^{-1} for monochromatic radiation of frequency 540 × 10^{12} Hz.	坎德拉（cd）：国际单位制中发光强度的基本单位，定义为频率540 × 10^{12}赫兹的单色辐射在每球面度上1/683瓦特的强度。
The **radian, rad**, is the SI derived unit for measuring angles. One radian is the angle subtended at the center of a circle by an arc that is equal in length to the radius of the circle.	弧度（rad）：国际单位制中角度的导出单位。1弧度定义为圆上长度等于圆半径的弧所对应的圆心角大小。
The **steradian, sr**, is the SI derived unit of solid angle. A steradian can be defined as the solid angle subtended at the center of a sphere of unit radius by a unit area on its surface.	球面度（sr）：国际单位制中立体角的导出单位。1球面度的定义为单位半径球体单位面积的球面在球心处对应的立体角。
The **Hertz, Hz**, is the SI derived unit of frequency and is defined as one cycle per second.	赫兹（Hz）：国际单位制中频率的导出单位，定义为每秒1个周期。
The **Newton, N**, is the SI derived unit of force. One newton is the force needed to accelerate one kilogram of mass at the rate of one meter per second squared in the direction of the applied force.	牛顿（N）：国际单位制中力的导出单位。1牛顿定义为使1千克的质量在受力方向上获得1米/秒²的加速度所需要的力。

PHYSICAL QUANTITIES

物理量

Definition	定义
The **Pascal**, **Pa**, is the SI derived unit of pressure. One pascal is the pressure exerted by a force of magnitude one newton perpendicularly upon an area of one square meter.	帕斯卡（Pa）：国际单位制中压强的导出单位。1帕斯卡等于大小为1牛顿的力垂直作用在1平方米面积上所产生的压强。
The **Joule**, **J**, is the SI derived unit of energy or work. It is equal to the energy transferred to (or work done on) an object when a force of one newton acts on that object in the direction of its motion through a distance of one meter.	焦耳（J）：国际单位制中能量或功的导出单位。1焦耳等于1牛顿的力作用在物体上，使其沿力的方向运动1米后，传递给物体的能量（或对物体做的功）。
The **Watt**, **W**, is the SI unit of power and it is defined as a derived unit of 1 joule per second. It is used to quantify the rate of energy transfer.	瓦特（W）：国际单位制中功率的导出单位。1瓦特定义为1焦耳/秒。瓦特用于量化能量传递的速率。
The **Coulomb**, **C**, is the SI derived unit of electric charge and it is defined as the quantity of electricity transported in one second by a current of one ampere.	库仑（C）：国际单位制中电荷的导出单位。1库仑定义为1秒内通过1安培电流传输的电量。
The **Volt**, **V**, is the SI derived unit for electric potential, electric potential difference (voltage), and electromotive force and it is defined as the potential difference between two points that will impart one joule of energy per coulomb of charge that passes through it.	伏特（V）：国际单位制中电势、电势差（电压）和电动势的导出单位。伏特定义为两点间的电势差，即1伏特等于使1库仑的电荷从一点运动到另一点做1焦耳功时所需要的电势差。

PHYSICAL QUANTITIES

物理量

Definition	定义
The **Farad, F**, is the SI derived unit of electrical capacitance, the ability of a body to store an electrical charge. One farad is defined as the capacitance which stores one coulomb of charge across a potential difference of one volt.	法拉（F）：国际单位制中电容的导出单位，电容是物体储存电荷的能力。1法拉定义为在1伏特的电势差上储存1库仑电荷的电容。
The **Ohm, Ω**, is the SI derived unit of electrical resistance and is defined as the electrical resistance between two points of a conductor when a constant potential difference of one volt, applied to these points, produces in the conductor a current of one ampere, the conductor not being a source of any electromotive force.	欧姆（Ω）：国际单位制中电阻的导出单位。1欧姆定义为在导体两点施加1伏特的恒定电势差，产生1安培的电流时对应的电阻，且该导体不能是电动势源。
The **Weber, Wb**, is the SI unit of magnetic flux. The Weber is the magnetic flux which, linking a circuit of one turn, would produce in it an electromotive force of 1 volt if it was reduced to zero at a uniform rate in 1 second.	韦伯（Wb）：国际单位制中磁通量的单位。1韦伯定义为穿过1个单匝导线圈的磁通量。它在1秒内匀速减小到零将产生1伏特电动势。
The **Tesla, T**, is the SI derived unit used to measure magnetic field strength. One tesla is defined as the field strength generating one newton of force per ampere of current per meter of conductor.	特斯拉（T）：国际单位制磁场强度的单位。1特斯拉定义为磁场中的导体在有1安培电流流过时，每米导体受到1牛顿的作用力所需的场强。
The **Henry, H**, is the SI derived unit of electrical inductance. If the rate of change of current in a circuit is one ampere per second and the resulting electromotive force is one volt, then the inductance of the circuit is one henry.	亨利（H）：国际单位制中电感的导出单位。如果电路中电流的变化率是每秒1安培，产生的电动势是1伏特，那么该电路的电感是1亨利。

SI BASE UNITS and DERIVED UNITS

国际单位制基本单位和导出单位

	Quantity	物理量	Name	名称	Symbol	符号
SI Base Units 国际单位制基本单位	Length	长度	Meter	米	m	m
	Mass	质量	Kilogram	千克	kg	kg
	Time	时间	Second	秒	s	s
	Electric current	电流	Ampere	安培	A	A
	Temperature	温度	Kelvin	开尔文	K	K
	Amount of substance	物质的量	Mole	摩尔	mol	mol
	Luminous intensity	发光强度	Candela	坎德拉	cd	cd
SI Derived Units 国际单位制导出单位	Plane angle	平面角	Radian	弧度	rad	rad
	Solid angle	立体角	Steradian	球面度	sr	sr
	Frequency	频率	Hertz	赫兹	Hz	Hz
	Force	力	Newton	牛顿	N	N
	Pressure	压强	Pascal	帕斯卡	Pa	Pa
	Energy Work Heat	能量 功 热	Joule	焦耳	J	J
	Power	功率	Watt	瓦特	W	W
	Electric charge	电荷	Coulomb	库仑	C	C
	Electric potential (emf)	电势 (电动势)	Volt	伏特	V	V
	Capacitance	电容	Farad	法拉	F	F
	Electric resistance	电阻	Ohm	欧姆	Ω	Ω
	Magnetic flux	磁通量	Weber	韦伯	Wb	Wb
	Magnetic field strength	磁场强度	Tesla	特斯拉	T	T
	Inductance	电感	Henry	亨利	H	H

SELECTED PREFIXES for POWERS of TEN

常用10的幂次前缀

Power/幂次	Name	名称	Symbol	符号
10^{24}	yotta	尧	Y	Y
10^{21}	zetta	泽	Z	Z
10^{18}	exa	艾	E	E
10^{15}	peta	拍	P	P
10^{12}	tera	太	T	T
10^{9}	giga	吉	G	G
10^{6}	mega	兆	M	M
10^{3}	kilo	千	k	k
10^{2}	hecto	百	h	h
10^{1}	deca	十	da	da
10^{-1}	deci	分	d	d
10^{-2}	centi	厘	c	c
10^{-3}	milli	毫	m	m
10^{-6}	micro	微	μ	μ
10^{-9}	nano	纳	n	n
10^{-12}	pico	皮	p	p
10^{-15}	femto	飞	f	f
10^{-18}	atto	阿	a	a
10^{-21}	zepto	仄	z	z
10^{-24}	yocto	幺	y	y

MATHEMATICAL SYMBOLS and OPERATIONS

数学符号和运算

Symbol/符号	Meaning	含义
$=$	is equal to	等于
$\neq$	is not equal to	不等于
$\approx$	is approximately equal to	约等于
$>$	is greater than	大于
$<$	is less than	小于
$\geq$	is greater than or equal to	大于或等于
$\leq$	is less than or equal to	小于或等于
$\gg (\ll)$	is much greater (less) than	远大于（小于）
$\sim$	weak approximation	弱近似
$\propto$	is proportional to	正比于
$\simeq$	is defined as	定义为
$\cdot$	dot (scalar product)	点乘（标量积）
$\times$	cross (vector product)	叉乘（矢量积）
Δx	the change in x	x的变化量
$\sum_{i=1}^{N}$	the sum of all quantities x_i from $i = 1$ to $i = N$	x_i从$i = 1$到$i = N$所有量的和
$\lvert x \rvert$	the magnitude of x (always a non-negative quantity)	x的绝对值（非负值）
$f(x)$	function of x	x的函数
$\Delta x \to 0$	Δx approaches zero	Δx趋向0
dx/dt	the derivative of x with respect to t	x对t求导
x'	derivative (Lagrange's notation)	导数（拉格朗日的符号）
$\partial x/\partial t$	the partial derivative of x with respect to t	x对t的偏导
$\int$	integral	积分
$\therefore$	therefore	因此
$\Rightarrow$	implies	推出

ALGEBRA

代数

Symbol/符号	Meaning	含义
x, y, z	unknowns or variables	未知量或变量
$ax = b$ $x + a = b$ $x/a = b$	equations	方程
x^a — exponent of x / x的指数	quantity x raised to the a-th power	x的a次方
$\dfrac{a}{b}$ — numerator 分子 / denominator 分母	fraction	分数
$ax + ay + az = a(x+y+z)$ common factor 公因数 $a^2 + 2ab + b^2 = (a+b)^2$ perfect square 完全平方 $a^2 - b^2 = (a+b)(a-b)$ difference of two squares 平方差	formulas for factoring an equation	方程因式分解的公式
$ax^2 + bx + c = 0$ a, b, c — coefficients of equation 方程的系数	quadratic equation	二次方程式
$y = mx + b$ slope of line 线的斜率	linear equation, or equation of a straight line	线性方程或直线方程
$y = \log_a x$ base 基数	logarithm of x with respect to base a	x关于a的对数
$x = \mathrm{antilog}_a y = a^y$	antilogarithm of y	y的真数
$a = 10$	common or decimal base of logarithms	对数的公制或十进制底
$e = 2.718\ldots$	natural base of logarithms	对数的自然底
$\pi = 3.141\,592\,6\ldots$	number pi	圆周率π

GEOMETRY

几何

Symbol/符号		Meaning	含义
$y = mx + c$		equation of a straight line	直线方程
$x^2 + y^2 = R^2$		equation of a circle of radius R	半径为R的圆的方程
$\dfrac{x^2}{a^2} + \dfrac{y^2}{b^2} = 1$		equation of an ellipse	椭圆方程
$y = ax^2 + b$		equation of a parabola	抛物线方程
$xy = \text{const}$		equation of a rectangular hyperbola	直角双曲线方程
	Area/面积 $= ab$	rectangle	矩形
	Area/面积 $= \pi r^2$ Circumference/周长 $= 2\pi r$	circle	圆形
	Area/面积 $= bh/2$	triangle	三角形
	Surface area/表面积 $= 4\pi r^2$ Volume/体积 $= 4\pi r^3/3$	sphere	球体
	Volume/体积 $= \pi r^2 l$	cylinder	圆柱体
	Volume/体积 $= abc$	parallelepiped	平行六面体

TRIGONOMETRY

三角函数

Symbol/符号	Meaning	含义
o = opposite side　θ的对边 a = adjacent side　θ的邻边 h = hypotenuse 　　直角三角形的斜边	right-angled triangle	直角三角形
	side opposite the angle	对角边
	side adjacent to the angle	与直角相邻的边
	hypotenuse	直角三角形的斜边
$\sin\theta = \dfrac{\text{side opposite }\theta\ (\theta\text{的对边})}{\text{hypotenuse }\ (\text{直角三角形的斜边})}$	sine	正弦
$\cos\theta = \dfrac{\text{side adjacent to }\theta\ (\theta\text{的邻边})}{\text{hypotenuse }\ (\text{直角三角形的斜边})}$	cosine	余弦
$\tan\theta = \dfrac{\text{side opposite }\theta\ (\theta\text{的对边})}{\text{side adjacent to }\theta\ (\theta\text{的邻边})}$	tangent	正切
$\csc\theta = \dfrac{1}{\sin\theta}$	cosecant	余割
$\sec\theta = \dfrac{1}{\cos\theta}$	secant	正割
$\cot\theta = \dfrac{1}{\tan\theta}$	cotangent	余切
$c^2 = a^2 + b^2$	Pythagorean theorem	勾股定理
$a^2 = b^2 + c^2 - 2bc\cos\alpha$ $b^2 = a^2 + c^2 - 2ac\cos\beta$ $c^2 = a^2 + b^2 - 2ab\cos\gamma$	Cosine rule	余弦定理
$\dfrac{a}{\sin\alpha} = \dfrac{b}{\sin\beta} = \dfrac{c}{\sin\gamma}$	Sine rule	正弦定理

INDEX
索引

ENGLISH INDEX - 314
英文索引

A

aberration 264
abscissa 20
absolute pressure 86
absolute temperature 150
absolute zero 114
absorbed thermal energy 144
AC circuit 232
AC circuit consisting of a capacitor 236
AC circuit consisting of an inductor 234
AC generator 232
AC voltage 240
acceleration 36, 48
acceleration due to gravity 38
acceleration of the center of mass 62
accommodation 266
accumulation of charge 196
action force 50
addition of vectors 24
adiabatic expansion 120
adiabatic process 120, 146
adiabatically compressed 148
adiabatically expanded 148
adjustable mirror 276
air film 274
air-fuel mixture 148
alcohol thermometer 114
aligned magnetic dipole moment 208
alternating current 232
alternating current generator (AC generator) 218
alternating emf 218
alternating emf varies sinusoidally with time 218
altimeter 134
altimetry 134
altitude 86, 134
amber 154
ammeter 188
amorphous material 288
amount of thermal energy 150
Ampère–Maxwell law 224
amplitude 94, 102
analyzer 286
angle 26
angle in degrees 68
angle in radians 68
angle of deviation 254, 280
angle of incidence 250, 252
angle of reflection 250
angle of refraction 252
angle of shear 84
angle resolution 278
angular acceleration 68
angular displacement 68
angular frequency 94, 102, 192
angular frequency of oscillation 228
angular magnification 266
angular momentum 76, 206
angular momentum of a particle 76
angular momentum of a system of particles 76
angular position 68
angular speed 68
angular wave number 102
anisotropic material 288
antinode 110
apex angle of prism 254
applied voltage varies sinusoidally 238
approximately hyperbolic curve 140
aqueous humor 266
arbitrarily shaped wire segment 194
arc length 68
Archimedes' principle 88
area of emitting body 128
area of plate 170
area of the circular orbit 206
area under a curve 52, 148
area vector 214
array of atoms 282
arrow over the letter 22
associative law of addition 24
assumption 140
astigmatism 264
asymmetric shape 136
atmospheric pressure 86, 148
atomic physics 14
atomic spacing 282
atoms (ions) with permanent magnetic
 dipole moments 210
attractive electric force 156
attractive force 78, 154
audible (sound) wave 106
average acceleration 36
average angular acceleration 68
average coefficient of linear expansion 116
average coefficient of volume expansion 116

average current 176
average distance 138
average power 54, 238
average speed 136
average value of current 232
average value of the square of velocity 130
average velocity 34
Avogadro's number 118
axis 20
axis of rotation 66, 72, 218
axis of symmetry 62

B

B versus H curve 208
ballistic pendulum 96
bar magnet 190
barometric distribution 134
battery 182, 226
battery connected to resistor 182
BCS theory 180
beam of light 250
beam of white light 254
beat 110
Bernoulli's equation 90
biconcave lens 262
biconvex lens 262
bimetallic strip 116
birefringent material 288
black body 128
black hole 78
blurred image 264
body 30
body at rest 46
body in motion 46
Bohr magneton 206
boiling 122
boiling point 114
Boltzmann's constant 118
boundary 162, 250
boundary between media 258
Boyle's law 120
Bragg's law 282
braking system 220
brass rod 116
Brewster's angle 290
bright circular fringe 274

bright fringe 268
brownian motion 30
bulk metal 220
bulk modulus 84
buoyant force 88

C

calcite 288
calorimeter 122
calorimetry 122
capacitance 228
capacitance of capacitor 170
capacitive reactance 236
capacitor 168, 170, 230
capacitor connected across terminals of
 AC generator 236
capacitor connected to inductor 228
capacitor with dielectric 170
carbon brush 218
Carnot cycle 146, 148
Carnot engine 146
Carnot's theorem 146
Cartesian coordinate system 20
cavity resonator 128
Celsius temperature scale 114
center of curvature 42
center of gravity 62
center of lens 264
center of mass 96
center of mass of the system 62
center of rotation 42
central force 44
central gravitational force 78
centrifugal force 44
centripetal acceleration 42, 192
centripetal force 42, 44
CGS (centimeter-gram-second) system 18
change in direction of velocity vector 42
change in energy 54, 124
change in entropy 150
change in length 116
change in momentum 60
change in position 32
change in quantity 22
change in speed 42

change in volume 116
change of magnetic flux 214
changing electric field 200
changing flux 220
changing magnetic field 216
changing magnetic flux 224
charge 154, 158
charge carrier 176, 196
charge conserved 154
charge density 158
charge distribution 158, 162, 222
charge enclosed by surface 162
charge inside surface 162
charge on a conductor 168
charge oscillation 228
charge pump 182
charge transfer process 172
charged capacitor 228
charged conductor 164
charged particle 156, 192
charged particles collide with atoms 212
charged particles originate from the Sun 212
charged particles spiral around the field lines 212
charged particles trapped (deflected) by the Earth's magnetic field 212
charges separated by a distance 174
charging of capacitor 230
Charles' law 120
chromatic aberration 264
circuit 178, 186, 230
circuit diagram 182
circular aperture 278
circular arc 66
circular fringe 274
circular loop 202
circular motion 30, 66
circular orbit 80, 206
circular path 42, 192, 200
circular path centered on the wire 200
circulating current 220
circumference 68
circumference of circle 200
classical model of an atom 206
classical physics 14
Clausius' statement of the second law 142
clockwise 20
clockwise motion 30

clockwise rotation 68
closed circuit 186, 214
closed circular path 200
closed current-carrying conductor 194
closed Gaussian surface 162
closed loop 232
closed path 224
closed surface 160, 222
closed volume 222
closely spaced slits 280
closely spaced turn of wire 202
CM = center of mass 74
coaxial 170
coefficient of damping (damping ratio) ζ 98
coefficient of performance (COP) 144
coefficient of resistance of medium 98
coefficient of viscosity 92
coherent source of light 268
coil 202, 214
coil of ring-shaped form 202
coil of wire wound around a core 240
cold reservoir 142
colliding particle 60
collision 60
collision frequency 138
column of mercury 86
coma 264
combination 172
combustion 148
commercial generator 218
commutative law of addition 24
commutative law of multiplication 26
commutator 218
compass needle 190
completely polarized light 290
complex molecule 132
complex wave 110
component 22
compressed (condensation) region 106
compressibility 82
compressible medium 106
compression 82, 146
compression stroke 148
concave mirror 260
concentric 170
concept 14
concurrent force 64

condensation 122
condensed matter physics 14
condition for constructive interference 268
condition for destructive interference 268
condition for intensity minima 278
condition for interference 268
condition of equilibrium 64
condition of rotational equilibrium 64
condition of translational equilibrium 64
conducting loop 216
conducting shell 170
conducting wire 172
conduction 154
conduction current 200, 224
conduction electron 176
conductivity 178
conductor 154, 164, 168, 178, 196
conductor in electrostatic equilibrium 164
conductor of finite size 198
cone 108
cone cell 266
confocal microscope 266
conical wave front 108
connected in parallel 188
connected in series 188
connecting wire 182
connection 184
conservation of angular momentum law 76
conservation of energy 56
conservative force 44, 56
conservative gravitational force 78
constant acceleration 36
constant of proportionality 82
constant phase relationship 268
constant velocity 34
constant-volume gas thermometer 114
constructive interference 104
contact force 44
contact point 74
continuous charge distribution 158
convection 126
converging lens 262
convex mirror 260
convex-concave lens 262
coordinate 20
coordinates of the center of mass 62

coplanar force 64
core of soft iron 240
cornea 266
corpuscular (particle) theory 248
cosine 26
cosine function 24
cosmic ray 212
Coulomb constant 156
counterclockwise 20
counterclockwise (or anticlockwise) motion 30
counterclockwise rotation 68
couple 64
crest wave 102
critical angle 258
critical pressure p_c 140
critical resistance 230
critical temperature 180
critical temperature T_c 140
critical volume V_c 140
critically damped oscillation 98
cross 194
cross product 22, 28
cross symbol 28
cross-sectional area 82
crystalline lens 266
crystalline material 288
crystalline plane 282
crystalline structure 282
crystalline substance 208
Curie temperature 208, 210
Curie's law 210
current (voltage) phasor 234, 236
current 176, 220, 230
current 90° out of phase with stored charge 228
current and voltage in phase 232
current and voltage out of phase 234
current and voltage vary sinusoidally 232
current clockwise 220
current counterclockwise 220
current decays exponentially 230
current decreases steadily 220
current density 176, 178
current element 198
current entering junction 186
current in conductor 176

current lags voltage by 90° 234
current leads voltage by 90° 236
current leaving junction 186
current loop 206
current oscillates sinusoidally 230
current oscillation 228
current rises to the equilibrium value 230
current versus time 228, 230
current with voltage 90° out of phase 236
current with voltage not in phase 236
current-carrying conductor 190, 194, 198
current-carrying loop 194
current-carrying wire 194
current-voltage relationship 178
curvature of field 264
curve of speed distribution 136
curved path 40, 42
curvilinear motion 30
cycle 142, 148
cyclic process 124, 142
cycloid path 74
cyclotron 192
cyclotron frequency 192
cylindrical capacitor 170
cylindrical coordinate system 20
cylindrical shell 70, 170
cylindrical symmetry 162

D

damped beat 110
damped harmonic motion 98
damped oscillator 98
damped simple harmonic motion 230
damped system 100
damping coefficient 100
damping rate 98
dark circular fringe 274
dark fringe 268
DC current 218
dee 192
deflection in one direction 214
deformation of solid 82
degree of alignment 174
degree of polarization 284
degree of freedom 132
demagnetized substance 208
density 86

density of the atmosphere 134
density of the charge carrier 196
density of the magnetic flux 204
deposition 122
destructive interference 104
deviation of light by a prism 254
Dewar flask 126
diagonal of parallelogram 24
diamagnetic 206
diamagnetic substance 210
diatomic molecule 132
dielectric 168
dielectric constant 170
dielectric material 170
dielectric strength 170
Diesel engine 148
difference in optical path length 272
diffracted beam 280
diffracted beams as a result of interference and diffraction 280
diffracted beams interfere with each other 280
diffracted beams produce final pattern 280
diffraction 278
diffraction angle 278
diffraction grating 280
diffraction pattern 278
diffuse reflection 250
digital multimeter 188
dimensionless number 252
diminished image 260
dipole 210
dipole induced in direction opposite to the applied field 210
dipoles line up with the field 210
direct current generator (DC generator) 218
directed along the line 156
directed radially outward 158
directed towards the charge 158
direction 22, 28, 34
direction of current 176
direction of displacement 52
direction of electric field vector 284
direction of force 72
direction of induced current 216
direction of induced emf 216
direction of polarization 284
direction of propagation 102

direction of the magnetic field 190
direction of wave propagation 284
direction toward loop 216
directly proportional 48
discharge 172
discharged capacitor 228
discharging of capacitor 230
discrete crystalline plane 282
disk 70
disorder 150
displaced volume of fluid 88
displacement 32, 52, 102
displacement amplitude 106
displacement from equilibrium 94
displacement vector 32
dissipative force 98
distance 32
distance of separation 156
distortion 264
distribution of molecules in the atmosphere 134
distributive law of multiplication 26
disturbance 102
disturbed medium 102
divergence of light 278
diverging lens 262
domain wall 208
domain 208
domain aligned with field 208
domain randomly oriented 208
domain reoriented 208
Doppler effect 106
dot 194
dot product 22, 26
dot symbol 26
double-refracting material 288
downward force 86
drag force 220
drift speed 176, 196
driving force 100
dual nature 248
dyne 48

E

Earth's gravity 38
Earth's magnetic field 212
Earth's magnetic field pattern 212
Earth's magnetic field source 212
Earth's nonuniform magnetic field 212
eddy current 220
edge effect 170
effective diameter 138
efficiency of a Carnot engine 146
efficiency of cycle 148
elastic collision 60
elastic limit 82
elastic modulus 82
elastic pendulum 96
elastic potential energy 54
elasticity 82
elasticity in length 84
elasticity of shape 84
electostatic shielding or Faraday cage 164
electric dipole 174
electric dipole moment 174
electric field 160
electric field in the presence of a dielectric 174
electric field line 158, 160
electric field of capacitor 228
electric field strength 158, 160
electric field vector 158, 222, 224
electric field without dielectric 174
electric flux 160, 224
electric flux entering surface 160
electric flux leaving surface 160
electric flux through surface 160
electric force 156, 158
electrically charged 154
electrically isolated conductor 168
electrified 154
electrolytic capacitor 170
electromagnetic force 44
electromagnetic induction 214
electromagnetic radiation 78
electromagnetic wave type 244
electromagnetic wave 242
electromagnetic waves carry energy and momentum 242
electromagnetic waves consist of oscillating electric and magnetic fields 242
electromagnetic waves exert pressure on a surface 242
electromagnetic wave generated by oscillating electric charges 242
electromagnetic wave of short wavelength 282

electromagnetic wave predicted by Maxwell's equations 242
electromagnetism 14
electrometer 154
electron 154
electron gas 176
electroscope 154
electrostatic force 156
electrostatic generator 164
electrostatics 154
elliptical orbit 80
elliptically polarized light wave 284
emf (ee-em-ef) 182
emf (electromotive force) 224
emf induced in inductor 234
emf of battery 230
emf traversed in the direction of the emf 186
emf traversed in the direction opposite to the emf 186
emissivity 128
empirical constant 140
energy 54
energy dissipated as Joule heat 228
energy loss 100
energy of photon 248
energy oscillation 228
energy quantization 132
energy stored in capacitor 228, 230
energy stored in inductor 228, 230
energy transfer 228
energy transport 78
engine 146, 148
entirely reflected light beam 258
entropy function 150
entropy function decreases 150
entropy function increases 150
envelope of oscillatory curve 98
envelope of wave 108
equal charges 174
equality of vectors 24
equally spaced slits 280
equation of continuity 90
equation of state 118
equilibrium state 150
equilibrium value of current 226
equipartition theorem 132
equipotential line 166

equipotential surface 166
equivalent capacitance 172
equivalent capacitor 172
equivalent collision 138
equivalent force 64
equivalent resistance 184
escape speed 80
evaporation 122
exhaust stroke 148
exhaust valve 148
expansion 146
expelled thermal energy 144
experiment 14
experimental curve 140
experimental observation 14
exponential factor 98
exponential form 134
external charge 158
external electric field 174
external force 44, 46, 58, 100
external magnetic flux 216
extraordinary ray 288
extreme value of charge 228
extreme value of current 228
eye focuses light 266
eye produces a sharp image 266
eyepiece 266

F

Fahrenheit temperature scale 114
Faraday's law of induction 220, 224
farsightedness (hyperopia) 266
ferromagnet 208
ferromagnetic 206
ferromagnetic substance 208
fiber optics 258
fictitious force 44
field force 44
field-ion microscope 166
field-ion microscope image 166
final equilibrium state 150
final length 116
final momentum 58
final position 32
final state of a system 124
final velocity 36
first (second...) harmonic 110

first-law equation 124
fixed axis 66
Fizeau's technique 248
flat (plane) mirror 260
flat glass surface 274
flat strip 196
flat surface 250
floating object 88
flow 90, 92
flow of charged particles 176
flow of charges 176
flow of electrons 176
flowtube 90
fluid 86, 88, 90
fluid in motion 88
focal length 260, 262
focal plane 264
focal point 80, 260
foot-pound-second system 18
force 44
force acting on an object 52
force caused by multiple charges 156
force couple 194
force per unit area 86
forced convection 126
forced oscillator 100
forms of radiation 244
Foucault current 220
four-stroke cycle 148
fraction of molecules with speeds between
    v and $v+dv$ 136
Fraunhofer diffraction 278
free electron 164
free-body diagram 48
free-fall 36
free-fall acceleration 38, 78
free-fall motion 36
freezing 122
freezing point 114
frequency 94, 102, 244
frequency of generator 218
Fresnel diffraction 278
friction between layers 92
friction force 44
frictionless pulley 48
fringe 268
fully charged capacitor 230

function of speed distribution 136
function of time 32
fundamental frequency 110
fundamental force 44
fundamental law 156

G

galvanometer 188, 214, 226
galvanometer needle 214
gamma ray 244
gamma ray emitted by radioactive nuclei
    during nuclear reaction 244
gas 118
gas at atmospheric pressure 118
gas at room temperature 118
gaseous state 140
gasoline engine 148
gauge pressure (overpressure) 86
Gaussian surface 162
Gauss's law 222
Gauss's law for magnetism 204, 222
Gay-Lussac's law 120
generalized form of Ampère's law 200
generator 218
geographic pole 212
geometric optics 248
geometrical arrangement of
    conductor 168
glancing collision 60
glass (plastic) rod 258
gradient operator 166
graph of charge versus time 228
grating 280
gravitation 38
gravitational field 38, 78
gravitational force 38, 44, 78, 96
gravitational mass 46
gravitational potential energy 54, 78, 134
gravitational radiation 78
gravitational wave 78
gravity 38
greater (high) index of refraction 258
greater than unity 252
Greek letter delta Δ 32
grounded conductor 154
gyroscope 76

H

half-apex angle 108
half of a period 110
half-silvered mirror 276
Hall coefficient 196
Hall effect current sensor 196
Hall field 196
Hall voltage 196
hard ferromagnet 208
head-on collision 60
heat 114, 122
heat capacity 122
heat conduction 126
heat death of the Universe 150
heat engine 142, 144, 146
heat loss 126, 146
heat loss by conduction 146
heat loss by friction 146
heat pump 144
helical path 192
helix form 202
high energy physics 14
high-frequency electromagnetic wave 248
high-Q (low-Q) circuit 238
high-temperature superconductor 180
high-voltage capacitor 170
hollow cylinder 70
hologram 290
holography 290
home insulation 126
hoop 70
horizontal component of velocity 40
horizontal direction 40
horizontal position 40
horizontal range 40
horseshoe magnet 190
hot reservoir 142
hydraulic lift 88
hydraulic press 88
hyperbola 120
hyperbolic curve 140
hypothesis 14
hysteresis loop 208

I

ice point of water 114
ideal absorber 128
ideal fluid 90
ideal gas 118, 130, 140
ideal gas law 118
ideal system 98
ignited 148
image 260, 262, 264
image distance 260
image in front (back) of mirror 260
imaginary particle 62
impedance 238
imperfect image 264
impulse of a force 58
impulse-momentum theorem 58
incident angle 258
incident light 250
incident (light) ray 250, 252, 290
inclined plane 48
incoherent sources of light 268
incompressible fluid 90
increasing altitude 134
independent of the path 166
index of refraction 252, 258
index of refraction decreases with increasing wavelength 254
individual resistance 184
induced charge 154
induced current 214, 220
induced emf 214, 220
induced moment 174
induced polarization 174
inductance 228
inductance of coil 226
induction 154
inductive circuit 234
inductive reactance 234
inductor 230, 234
inductor connected to terminals of AC generator 234
inelastic collision 60
inertia 46, 70
inertial frame 46
inertial mass 46
infinite resistance 188
infinitesimal process 150
infrared thermometer 114
infrared wave (IR) or heat wave 244
infrasonic wave 106

initial charge 228
initial equilibrium state 150
initial length 116
initial momentum 58
initial position 32
initial state of a system 124
initial velocity 36
instantaneous velocity 34
instantaneous acceleration 36
instantaneous angular acceleration 68
instantaneous current 176
instantaneous power 54
instantaneous voltage drop 232
instantaneous voltage drop across capacitor 236
instantaneous voltage drop across inductor 234
insulating wall 126
insulator 154, 168
intake stroke 148
intake valve 148
integral over a closed surface 160, 222
integrand 198
intensity distribution 278
intensity distribution of double-slit 270
intensity of transmitted light 286
intensity of polarized light 286
intensity of polarized wave 286
intensity of wave 106
interface 258
interference 104
interference pattern 104, 268, 270
intermediate state of a system 124
internal combustion engine 148
internal electric field 174
internal energy 124
internal force 44, 58
internal friction or viscosity 92
internal reflection 258
internal resistance 182
International System of Units 18
intrinsic magnetization 206
inversely proportional 48, 156
inverse-square law 78
inverted image 260
IR generated by hot body and molecule 244
iris 266
iron filing 190
irregular flow 92
irregularly shaped conductor 164

irreversible (real) process 142, 150
irrotational flow 92
isentropic process 150
isobaric process 120
isolated conductor 164
isolated magnetic pole 204
isolated system 56, 58, 76, 124, 150
isotherm 120, 140
isothermal process 120, 146
isovolumetric (isochoric) process 120

J

junction 184, 186
junction rule 186

K

Kelvin temperature scale 114
Kelvin-Planck's statement of the second law 142
kilogram (kg) 18
kinematics 30
kinetic energy 54
kinetic friction force 44

L

label on the axis 20
laminar flow 90
latent heat 122
latent heat of fusion 122
latent heat of vaporization 122
lateral magnification 260, 266
Laue pattern 282
law 14, 120
law of action and reaction 50
law of conservation of momentum 58
law of heat conduction 126
law of inertia 46
law of motion 46
layers of fluid 92
LC circuit 228
left coil connected to input ac voltage source 240
length of a path 32
length of string 96
length of the wire element 198
lens 262
lens equation 262
lens thickest at the edge 262
lens thickest in the middle 262

lever arm 72
light 248, 290
light beam 248, 258, 276
light beam split into two rays 276
light intensity 248
light of wavelength 274
light ray 248, 250, 252
light ray bent at boundary 252
light ray encounters a boundary 252
light ray enters medium 252
light string 96
light wave 284
lightbulb 184
LIGO Laser Interferometer Gravitational-Wave Observatory 276
like charges 154
limit 36
limit of the ratio 34
line integral 200, 224
line integral over a path 224
line of action of a force 64, 72
linear charge density 158
linear current-voltage relationship 178
linear density 70
linear momentum 58
linear motion 30
linear wave 104
linearly polarized electromagnetic wave 242
linearly polarized light wave 284
liquefied gas 140
liquid 86
liquid nitrogen 180
liquid state 140
liquid-vapor equilibrium state 140
load resistance 182
location in space 20
logarithmic decrement 98
long thin rod 70
long wire 198, 202
longitudinal wave 102, 106
loop 186, 216, 226
loop of wire 214
loop of wire of a given area 214
loop rotation clockwise 194
loop rotation counterclockwise 194
loop rule 186
loosely (tightly) wound solenoid 202

Lorentz force 192
loss of energy 56
loss of kinetic energy 56
loud explosion 108
low-density gas 118
lower crystalline plane 282
lower index of refraction 258
low-pressure (rarefaction) region 106

M

Mach number 108
macroscopic state of a system 124
magnet 190
magnet held stationary relative to the loop 214
magnet moves away from the loop 214
magnet moves towards the loop 214
magnetic bottle 192
magnetic dipole moment 208
magnetic field 190
magnetic field at a point 198
magnetic field density 190
magnetic field line 190, 202
magnetic field lines converge at opposite end 202
magnetic field lines diverge from one end 202
magnetic field of inductor 228
magnetic field strength 206
magnetic field vector 222
magnetic flux 204, 224
magnetic flux changes in time 226
magnetic flux counteracts the increase of current 216
magnetic flux induces emf 226
magnetic flux produced by current 216
magnetic flux threads the circuit 214
magnetic flux through a closed surface 204
magnetic flux through an area element 204
magnetic flux through loop 216, 226
magnetic flux through one coil passes through the other coil 240
magnetic force 192
magnetic hysteresis 208
magnetic induction 190
magnetic moment 206
magnetic moment directed out of (into) page 194
magnetic monopole 222
magnetic permeability 206
magnetic pole 212, 218

magnetic susceptibility 206
magnetization 206
magnetization curve 208
magnetization of a substance 206
magnetization vector 206
magnetized substance 208
magnetometer 196
magnification 266
magnified image 260
magnitude 22, 26, 34
magnitude of charge 156
magnitude of the electric force 156
magnitude of torque 72
magnitude or strength of magnetic field 190
mass 48
mass attached to a spring 94
mass per unit volume 86
mass spectrometer 192
massless string 48
material separating the charged conductors 168
matrix algebra 186
maximum current 232
maximum height 40
maximum of mth order 280
maximum value of current 234
maximum value of voltage 234
[Maxwell-]Boltzmann distribution function 136
mean distance 138
mean free path 138
mean free time 138
mechanical system 96
mechanical wave 102
mechanics 14
medium 250, 252
melting 122
membrane 126
mercury barometer 86
mercury-in-glass thermometer 114
merging black holes 78
merry-go-round 76
metal 180
meter (m) 18
method for observing interference of light waves 274
microscopic model 130
microscopic region 208
microscopic state of a system 124
microwave 244

microwave generated by electronic device 244
minute detail 266
mirror 260, 276
mirror equation 260
mirror-like surface 250
modern physics 14
modes of vibration 110
modified equation of state 140
molar mass 118
molar specific heat 122
moment arm 72, 194
moment of inertia 70
moment of inertia of a rigid obect 70
moment of inertia of a system of particles 70
momentum 58, 76
monatomic ideal gas 130
monatomic molecule 132
monochromatic sources of light 268
monopole 204
Moon 80
most efficient engine 146
most probable speed 136
motion 30
motion along a straight line 30
motion in a plane 30
motion of charges 164
motion of molecule 132
motion of planets 80
moving charged particle 190, 192
moving object 30
moving through a magnetic field 220
multiloop circuit 186
multiple internal reflection 258
multiplication of a vector by a scalar 24
mutual inductance 226
mutual induction 226

N

natural convection 126
natural frequency 98, 100
nearly circular orbit 80
nearsightedness (myopia) 266
negative acceleration 36
negative charge 154
negative direction 20
negative displacement 32
negative electric flux 160

negative focal length 262
negative of a vector 24
negative terminal of battery 172
negative velocity 34
negligible intermolecular force 140
negligible of edge effect 170
negligible volume occupied by molecule 140
net charge 162, 222
net electric flux 160, 162, 222
net electric flux through a closed surface 162
net external torque 76
net force 44, 48
net magnetic field 202
net magnetic flux 204, 222
net work done 142
neutral equilibria 64
newton 48
Newton's Second Law for a system of particles 62
Newton's Theory of gravitation 38
no phase change 272
node 110
nonconservative force 44, 56
nondeformable object 66
nonlinear current-voltage relationship 178
nonlinear curve 140
nonlinear wave 104
nonohmic material 178
nonpolar molecule 174
nonspherical Gaussian surface 162
nonsteady or unsteady flow 92
nonuniform magnetic field 192
nonviscous fluid 90
normal 250
normal force 44, 50, 86
normal mode 110
normal to surface 160
north pole 190, 212
Northern Lights 212
N-turn coil 226
nuclear physics 14
number 22
number of collisions 138
number of degrees of freedom 132
number of independent equations 186
number of lines penetrating surface 160
number of lines per unit area 160
number of magnetic field lines entering surface 204
number of magnetic field lines leaving surface 204
number of molecules per unit volume 130
number of moles 118
number of oscillations 94
number of turns of wire 202
number of turns of wire per unit length 202

O

object (image) focal point 262
object 30, 262
object distance 260
object in equilibrium 64
objective 266
observer 106
Oersted's experiment 190
ohmic material 178
one quarter of a period 110
one-dimensional motion 30
open circuit 230
open to atmosphere 86
open-circuit voltage 182
open-tube manometer 86
opposite charges 174
oppositely charged conductors 168
optic(al) axis 260, 264, 288
optical fiber 258
optics 14
orbit 80
orbital angular momentum 206
orbital magnetic moment 206
orbital motion 206
orbital period 80
orbital speed 80
orbiting electron 206
order number of fringe 268
ordinary ray 288
ordinate 20
origin 20
oscillation 30, 94, 98
oscillation period 234
oscillator 98
oscillatory motion 30
Otto cycle 148
output voltage 240
overdamped oscillation 98
own magnetic flux 216

P

pair of two coherent light sources 270
pairwise interaction 50
parabola 40
parabolic mirror 266
parabolic path 40
parallel combination 172
parallel connection 184
parallel slits 280
parallel-axis theorem 70
parallelogram area 28
parallelogram rule of addition 24
parallel-plate capacitor 170
paramagnetic 206
paramagnetic substance 210
paraxial ray 260
partial derivative 166
partially aligned electric dipoles 174
partially polarized light 290
partially submerged object 88
particle 30, 156
particle model 30
particle-like behavior 248
Pascal's law 88
path 32, 40
path difference 268, 270, 272, 280
path integral or line integral 166
path length 110
peak of trajectory 40
pendulum 96
perfect absorber 126
perfect emitter 126
perfectly elastic collision 60
perfectly inelastic collision 60
perfectly reflecting surface 258
period 94, 102
period of rotation 192
periodic (oscillatory) motion 94
periodic force 100
permanent magnet 190
permanent magnetic dipole moment 208
permanent polarization 174
permeability of free space 198, 224
permittivity of free space 156, 162, 222, 224
perpendicular 28
phase 94
phase angle 104

phase angle between current and voltage 238
phase change 122
phase change of 180°   272
phase changes upon reflection 272
phase constant (or phase angle) 94, 102
phasor 232
phasor diagram 232, 234, 236
photon 248
physical (compound) pendulum 96
physical contact 60
physical phenomenon 14
physical quantity 14
physics 14
physics of elementary particles 14
pipe 90
piston 86, 148
pivot point 72, 96, 220
planar symmetry 162
Planck's constant 248
Planck's radiation law 128
plane 160
plane electromagnetic wave 242
plane formed by vectors 28
plane of incidence 250
plane of loop 194
plane of polarization 284
plane of symmetry 62
plane wave 102, 256
plane-polarized light wave 284
plano-concave lens 262
plano-convex lens 262, 274
plate separation 170
point charge 156, 158
point charges of opposite sign 156
point charges of same sign 156
point in a plane 20
point in space 20
point mass 96, 138
point of aplication of a force 64, 72
point on a line 20
point on the rim 74
point source 256
point soure of light 260
Poiseuille's equation 92
polar axis 20
polar coordinate system 20
polar molecule 174

polarity 196, 216
polarization 174
polarization by double refraction 284
polarization by reflection 284
polarization by scattering 284
polarization by selective absorption 284
polarized light 290
polarized light wave 284
polarizer 286
polarizing angle 290
polaroid 284
pole 190
polyatomic ideal gas 130
polygon method of addition 24
position 20, 32
position of antinode 110
position of node 110
position of the center of mass 62
position vector 32
position vector about origin 76
position-time graph 32, 34
positive acceleration 36
positive charge 154
positive direction 20
positive displacement 32
positive electric flux 160
positive focal length 262
positive susceptibility 210
positive terminal 188
positive terminal of battery 172
positive test charge 158
positive velocity 34
positive work 100
potential 166
potential difference (voltage) 166, 178
potential difference (voltage) across capacitor 172
potential difference across battery 182
potential difference across resistor 184
potential difference between conductors 168
potential drop 184, 186
potential energy 54, 166
potential of a charged conductor 166
potential of a pair of point charges 166
potential of a point charge 166
potentiometer 188
power 54
power factor 238

power of a lens 266
power stroke 148
Poynting vector 242
precise measurement of displacement 276
pressure 86, 118
pressure amplitude 106
pressure variation 108
pressure variation with altitude 134
primary point source or oscillator 256
primary winding (turns) 240
principal axis 260
principal maximum 280
prism 254
prism spectroscope 254
probability 136, 150
process 120, 142
projectile 40
projectile motion 30
projection 22
propagation of light (wave) 256
proportional 156
pulsating DC current 218
pupil 266
pure rotational motion 66
pure superconductor 180
Pythagoras theorem 24

Q

quality factor 98, 238
quantitative measurement 14
quantized charge 154
quantum Hall effect 196
quantum mechanics 14

R

radial acceleration 42
radiation 126
radio wave 244
radius (center) of curvature 260
radius of arc 68
radius of circle 68
radius of circular fringe 274
radius of curvature 42, 274
radius of curvature of lens surface 262
radius of path 192
radius of the circular orbit 206
random direction 250

random motion 30
randomly oriented dipoles 210
randomly oriented electric dipoles 174
random-walk process 138
range from v to $v+dv$ 136
rate of change of flux 224
ratio of stress to strain 82
ray approximation 248
ray diagram 260
ray path 256
ray reflected vertically upward 276
ray transmitted horizontally toward mirror 276
Rayleigh's criterion 278
reaction force 50
real fluid 90
real gas 130, 140
real image 260
real system 98
reconstructed wavefront 256
rectangular coordinate system 20
rectangular loop 194
rectangular plate 70
red light deviates the least 254
reference axis 20, 72
reference frame 20
reference line 68
reference point 20
reflected angle 258
reflected (light) ray 250, 252, 276, 290
reflecting surface 250
reflecting telescope 266
reflection 250
reflection grating 280
refracted angle 258
refracted (light) ray 252, 290
refracting telescope 266
refraction of light by a prism 254
refractive index 264
refrigerator 144
relative acceleration 36
relative motion 30
relative velocity 34
relativistic kinetic energy 54
remanent magnetization 208
repulsive electric force 156
repulsive force 154
residual gases 148

resistance 178, 182, 184, 188
resistanceless LC circuit 228
resistive force, drag 44
resistivity 178
resistivity approaches a finite value 180
resistivity decreases 180
resistivity drops to zero 180
resistivity increases 180
resistivity varies linearly 180
resistivity versus temperature 180
resistor 178, 230, 232
resistor traversed in the direction
 of the current 186
resistor traversed in the direction opposite
 to the current 186
resolving power 280
resonance 100
resonance curve 100
resonance frequency 100, 238
resonant vibration 100
restoring force 44, 96, 98
resultant external force 64
resultant external torque 64, 76
resultant force 44
resultant vector 24
resultant wave function 104
retarding force 98
retina 266
reversal of the magnetic field direction 212
reversible pendulum 96
reversible process 142, 146, 150
revolution 68, 192
right coil connected to load resistor 240
right-hand rule 28, 192, 200
rigid object 66
rigid-object model 66
RMS current 232
RMS speed 136
RMS voltage 232
rocket propulsion 50
rod cell 266
Roemer's method 248
rolling body 74
rolling down 74
rolling friction 74
rolling motion 30, 74
rolling on rough surface 74

rolling without slipping 74
root mean square speed (RMS) 130
rotating charged particle 192
rotating tendency 72
rotation about center of mass (CM) 66
rotation axis 70
rotation axis through center 70
rotation axis through end 70
rotational energy 70
rotational kinematics equation 66
rotational kinetic energy 74
rotational motion 30
rotational motion of molecule 132
rough surface 250
rubbed with silk or fur 154
rubber or glass rod 154
ruling (groove) 280

S

same plane 250
saturation 208
scalar product 22, 26
scalar projection (scalar component) 26
scalar quantity 22, 34
scale 20
screen with two narrow parallel slits 270
sea level 134
second (s) 18
secondary point source or oscillator 256
secondary wave 256
secondary winding (turns) 240
self-induced emf 226
self-induction 226
semiconductor 154, 180, 196
semimajor axis 80
series combination 172
series connection 184
series of colors 254
series RL circuit 230
series RLC circuit 230
set of axes 20
set of coordinates 20
set of streamlines 90
set up of induced current 214
shadowed region 278
sharp image 264
shear 82

shear modulus 84
shear strain 84
shear stress 84
shock wave 108
short circuit 170
short-range force 130
shunt 188
SI unit 18
SI-Système Internationale 18
side of triangle 24
sign of the charge carrier 196
simple harmonic motion 30
simple harmonic oscillator 94
simple magnifier 266
simple pendulum 96
sine function 24
single index of refraction 288
single wavelength 254
single-loop circuit 186
single-slit diffraction 278
single-wavelength light source 270
sinusoidal wave 102
slip ring 218
slit spacing 280
slit 280
slope 32
slope of line 34
slope of the graph 36
small susceptibility 210
smooth path 90, 92
smooth surface 250
Snell's law 252
soft ferromagnet 208
solar constant 128
solenoid 202
solenoid of finite size 202
solid cylinder 70
solid sphere 70
sonic boom 108
source 106
source of disturbance 102
source of emf 182
source of energy 182
source of light 248, 268
source speed 108
south pole 190, 212
space and time 30

spark plug 148
spatial pinhole 266
specific heat 122
spectrum 254
specular reflection 250
speed 34
speed distribution 136
speed distribution curve reaches a peak 136
speed gradient 92
speed of light 248, 252
speed of light in a medium 252
speed of light in a vacuum 252
speed of sound wave 106
spherical aberration 264
spherical capacitor 170
spherical coordinate system 20
spherical Gaussian surface 162
spherical mirror 260
spherical molecules collision 138
spherical shell 170
spherical symmetry 162
spherical wave 102, 256
spherical wave front 108
spin 206
spin angular momentum 206
spin magnetic moment 206
spring pendulum 96
square of the separation 156
standard of length 18
standard of mass 18
standard of time 18
state function 124, 150
state of a system 124
static friction force 44
stationary loop 216
stationary particle 156
steady current 176, 198, 200, 218
steady flow 90
steady magnetic field 190
steady-state circuit 230
steady-state conditions 100
steam engine 148
steam point of water 114
steel rod 116
Stefan-Boltzmann constant 128
Stefan-Boltzmann law 128
step-down transformer 240

step-up transformer 240
Stirling engine 148
stored charge 168
straight wire 198
straight wire segment 194
straight-line pass 278
strain 82, 84
streamline 90
stress 82, 84
strong magnetic field 202
strong nuclear force 44
sublimation 122
submerged object 88
subsonic speed 108
substance 208, 210
subtraction of vectors 24
sum up 198
Sun 80
superconducting magnet 180
superconductivity 180
superposition of potentials 166
superposition principle 104, 158, 268
supersonic speed 108
surface 250
surface area bounded by a path 224
surface bounded by closed path 200
surface charge 174
surface charge density 158
surface density 70
surface integral 204
surface of area A 160, 204
surface of cone 108
surface of cross-sectional area 176
surface perpendicular to direction of current 176
surface tangent to wavelet 256
surroundings 126
susceptibility 210
switch 226, 230
switch closed 228
symmetric charge distribution 162
symmetric object 62
system of lens 262
system of particles 62

T

tail 22
tail of arrow 194

tangent line 34
tangent to path 42
tangential acceleration 42
tangential force 84
technique for polarizing light 286
telecommunications 258
telescope 266
temperature 118
temperature coefficient of resistivity 180
temperature drops from T_1 to T_2 148
temperature gradient 126
temperature increases 148
temperature rises rapidly 148
temperature scale 114
temperature variations 116
tensile strain 84
tensile stress 84
tension 82
tension force 44
terminal of battery 172
terminal voltage 182
test charge 158
test charge at rest 158
test charge magnitude 158
testing of optical lenses 274
the 2nd Law of Thermodynamics statement 142
the Boltzmann distribution 134
theoretical engine 146
theory 14
theory of general relativity 78
theory of relativity 14
thermal conductivity 126
thermal contact 114
thermal efficiency 144
thermal energy 122, 142, 144
thermal energy absorption 142
thermal energy transfer 122
thermal equilibrium 114
thermal expansion joints 116
thermodynamic temperature scale 114
thermodynamic variables 118
thermodynamics 14
thermometer 114
thermostat 116
thin film 272
thin film of index of refraction 272
thin film of uniform thickness 272

thin spherical shell 70
three-dimensional diffraction grating 282
thrust 50
time constant 230
time interval 32, 36
time rate of change of angular momentum 76
time rate of change of current 230
tip 22
tip of arrow 194
top 76
toroid 202
torque (moment of force) 72, 174
torque about any point 194
torque on current loop 194
torsional pendulum 96
total acceleration 42
total charge 172
total current 200
total electric flux 222
total internal reflection 258
total kinetic energy 74
total magnetic field 198
total magnetic flux 204
total mass of the system 62
total mechanical energy 54
total momentum 58, 60
total momentum of a system of particles 62
total number of turns of wire 202
total work 52
totally submerged object 88
train slows down 220
trajectory 40
transformation of energy 56
transformer 240
translation of center of mass 66
translation of CM 74
translational kinetic energy 74
translational motion 30
translational motion of molecule 132
transmission (polarizer) axis 286
transmission grating 280
transmission line 240
transparent medium 252
transparent rod 258
transparent slits 280
transverse electromagnetic wave 242
transverse wave 102

travelling wave 102, 104
triangle method of addition 24
triple point 114
trough 102
turbulent flow 92
turn of wire 202
twisted loop 194
two indices of refraction 288
two rays recombine to produce an interference pattern 276
two rays travel separate paths 276
two-dimensional collision 60
two-dimensional motion 30

U

ultimate breaking limit 82
ultrasonic wave 106
ultraviolet light (UV) 244
ultraviolet light generated by the Sun 244
unaided (naked) eye 266
uncharged capacitor 172, 230
undamped oscillator 98
underdamped oscillation 98
uniform electric field 160
uniform electric field in direction 160
uniform electric field in magnitude 160
uniform gravitational field with air resistance 38
uniform gravitational field without air resistance 38
uniform magnetic field 190, 192, 202
unit vector 22, 26
unit vector i, j, k 20
unit 22
universal gas constant 118
universal gravitational constant 78
unlike charges 154
unmagnetized substance 208
unmagnified image 260
unpolarized (natural) light wave 284
unpolarized light 290
unstable (stable) equilibria 64
upper (lower) edge of the flat strip 196
upper crystalline plane 282
upright image 260
upward force 86

V

vacuum 126
Van Allen radiation belt 212
van de Graaff generator 164
variable capacitor 170
variation in pressure 106
variation of pressure with depth 86
varying current 230
vector position of the center of mass 62
vector product 22, 28
vector quantity 22, 34
velocity 34
velocity of the center of mass 62
velocity selector 192
velocity-time graph 36
vertical component of velocity 40
vertical position 40
very distant object 266
vibrational motion of molecule 132
vibrational (or oscillatory) motion 30
viewer (observer) 260
violet light deviates the most 254
virtual image 260
viscous fluid 92
visible light 212, 244
visible light produced by rearrangement of electrons in atoms and molecules 244
visible spark 172
visible wavelength 254
voltage across capacitor lags behind the current by 90° 238
voltage across inductor leads the current by 90° 238
voltage across resistor in phase with current 238
voltage drop 232
voltage drop across resistor 232
voltage stepped down 240
voltage stepped up 240
voltmeter 188
volume 118
volume charge density 158
volume density 70
volume elasticity 84
volume flux (flow rate) 90
volume strain 84
volume stress 84

W

water / water vapor / ice 114
wave 102, 106, 256
wave combination 104
wave equation 102
wave front 102, 108
wave function 102
wave function for stationary wave 110
wave incident on analyzer 286
wave motion 30
wave speed 102, 108, 110
wave theory 248
waveform 104
wavefront 256
wavefront position 256
wavelength 102, 244
wavelength of light 264
wavelet 256
waves passing around obstacles 278
waves passing through small openings 278
weak magnetic field 202
weak nuclear force 44
weight 38, 46, 78
weight of air 134
weightlessness 38
Wheatstone bridge 188
whirlpool region 92
wide range 244
Wien displacement law 128
wire 178
wire element 198
wire loop 218
wire segment 194
work done by a constant force 52
work done by a force 52
work done by a gas 124, 148
work done by a spring 52
work done by a varying force 52
work done per unit charge 182
work-energy theorem 54
working substance 142

X

X-axis 20
X-ray 244, 282
X-ray beam 282
X-rays generated by deceleration of
    high-energy electron 244

Y

Y-axis 20
Young's modulus 84

Z

Z-axis 20
zero acceleration 36
zero beat 110
zero displacement 32
zero resistance 188
zero velocity 34
zeroth law of thermodynamics 114

CHINESE INDEX – 336
中文索引

A

阿伏伽德罗常数 118
阿基米德定律 88
安培-麦克斯韦定律 224
暗条纹 268
暗圆条纹 274
凹槽 280
凹面镜 260
奥斯特实验 190
奥托循环 148

B

白光光束 254
摆 96
摆线路径 74
半导体 154, 180, 196
半顶角 108
半涂银镜 276
半周期 110
薄膜 126, 272
薄膜折射率 272
薄球壳 70
饱和 208
保守力 44, 56
保守引力 78
北极 190, 212
北极光 212
被地球磁场困住（偏转）的带电粒子 212
被干扰介质 102
被积函数 198
比1大 252
比辐射率/发射率 128
比例系数 82
比热 122
比值的极限 34
毕达哥拉斯定理 24
闭合的圆形路径 200
闭合电路 186, 214
闭合回路 232
闭合通电导体 194
边界 162, 250
边缘上的点 74
边缘效应 170
边缘效应可忽略不计 170
边缘最厚的镜片 262
变化磁场 216
变化的磁通量 220, 224
变化的电流 230
变化电场 200
变力做功 52
变压器 240
标量 22, 34
标量积 22, 26
标量投影 26
表面 250
表面密度 70
表压（过压） 86
冰点 114
并联 172, 184, 188
波 102, 106, 256
波长 102, 244
波的包络面 108
波的传播方向 284
波的混合 104
波的强度 106
波动 30
波动方程 102
波动理论 248
波峰 102
波腹 110
波腹位置 110
波谷 102
波函数 102
波节 110
波节位置 110
波前 102, 108, 256
波前位置 256
波速 102, 108, 110
波形 104
波义耳定律 120
波源 106
波源速度 108
玻尔磁子 206
玻尔兹曼常数 118
玻尔兹曼分布 134
玻璃（塑料）棒 258
伯努利方程式 90
不对称的形状 136
不发生形变的物体 66
不规则流 92
不可逆过程 142, 150
不可压缩流体 90
不稳定（稳定）平衡 64
布拉格定律 282

布朗运动　30
布儒斯特角　290
部分浸没的物体　88
部分偏振光　290
部分平行的电偶极子　174

C

擦边碰撞　60
参考点　20
参考系　20
参考线　68
参考轴　20, 72
测高法　134
层间摩擦　92
层流　90
叉乘　22, 194
叉乘符号　28
叉积　28
查理定律　120
柴油发动机　148
长半轴　80
长导线　202
长度变化　116
长度标准　18
长度弹性　84
场离子显微镜　166
场离子显微镜图像　166
场力　44
超导磁体　180
超导电性　180
超声波　106
超声速　108
成反比的　48
成正比的　48, 156
-成像清晰　266
呈正弦变化的电流和电压　232
乘法分配律　26
乘法交换律　26
冲击波　108
冲量-动量定理　58
充满电的电容器　230
重构波前　256
重新排列的磁畴　208
初动量　58
初级绕组　240
初始长度　116
初始平衡状态　150

初始位置　32
初速度　36
处于平衡状态的物体　64
储存的电荷　168
穿过曲面的线数　160
穿过小孔的波　278
传播方向　102
（电）传导　154
传导电流　200
传导热损失　146
传统发动机　148
串联　172, 184, 188
垂直的　28
垂直位置　40
垂直向上反射的光线　276
垂直于电流方向的面　176
垂直于曲面　160
纯超导体　180
纯转动　66
磁场　190
磁场大小或磁场强度　190
磁场的高斯定律　204, 222
磁场方向　190
磁场方向的翻转　212
磁场力　192
磁场密度　190
磁场强度　206
磁场矢量　222
磁场线　190, 202
磁场线在另一端聚拢　202
磁场线在一端发散　202
磁畴　208
磁畴壁　208
磁单极子　204
磁导率　206
磁感应强度　190
磁化率　206, 210
磁化强度　206
磁化强度曲线　208
磁化强度矢量　206
磁化物质　208
磁极　212, 218
磁矩　206
磁力计　196
磁偶极矩　208
磁瓶　192
磁体　190

磁体靠近线圈 214
磁体远离线圈 214
磁铁相对于线圈保持静止 214
磁通量 204, 224
磁通量变化 214
磁通量产生电动势 226
磁通量穿过电路 214
磁通量抵消电流的增加 216
磁通量密度 204
磁通量随时间变化 226
磁滞 208
磁滞回线 208
次级波 256
次级点源或振子 256
次级绕组 240
次声波 106
粗糙表面 250
存在介质时的电场 174

D

达因 48
大气密度 134
大气压 86, 148
大气压下的气体 118
大气中分子的分布 134
大小 22, 26, 34
大小均匀的电场 160
带电导体 164
带电导体的电势 166
带电的 154
带电电容器 228
带电粒子 156, 192
带电粒子和原子发生碰撞 212
带电粒子来自太阳 212
带电粒子流 176
带电粒子围绕磁场线旋转 212
带相反电荷的导体 168
单摆 96
单波长 254
单缝衍射 278
单个电阻 184
单回路电路 186
单色光源 268, 270
单位 22
单位长度的导线总匝数 202
单位电荷做的功 182
单位面积的线数 160

单位面积受力 86
单位矢量 22, 26
单位矢量 i、j、k 20
单位体积分子数 130
单位体积质量 86
单原子分子 132
单原子理想气体 130
单折射率 288
弹道摆 96
导电电子 176
导电壳 170
导体 154, 164, 168, 178, 196
导体的几何排布 168
导体间电势差 168
导体上的电荷 168
导体中的电流 176
导线 172, 178
导线圈 216
导线元 198
导线元的长度 198
导线匝数 202
导线总匝数 202
倒像 260
地理磁极 212
地球磁场 212
地球的磁场图谱 212
地球的磁场源 212
地球的非均匀磁场 212
地球引力 38
灯泡 184
等厚薄膜 272
等价力 64
等距狭缝 280
等量电荷 174
等容过程 120
等熵过程 150
等势面 166
等势线 166
等温过程 120
等温线 120, 140
等效电容 172
等效电容器 172
等效电阻 184
等压过程 120
低密度气体 118
低温热源 142
低压区域（疏部） 106

低折射率　258
笛卡儿坐标系　20
第m级极大　280
第二定律的开尔文-普朗克表述　142
第二定律的克劳修斯表述　142
第一定律方程　124
点乘　22, 194
点乘符号　26
点电荷　156, 158
点电荷的电势　166
点光源　260
点火　148
点积　26
点源　256
电场　160
电场强度　158, 160
电场矢量　158, 222, 224
电场矢量方向　284
电场线　158, 160
电池　182, 226
电池电动势　230
电池负极　172
电池间电势差　182
电池接线柱　172
电池正极　172
电磁波　242
电磁波的类型　244
电磁波对表面施加压力　242
电磁波具有能量和动能　242
电磁波由振荡电场和磁场组成　242
电磁辐射　78
电磁感应　154, 214
电磁力　44
电磁学　14
电导率　178
电动势　182, 224
电动势来源　182
电感　228
电感电路　234
电感器　228, 230
电感器储存的能量　228, 230
电感器的磁场　228
电感器的电压超前电流90°　238
电感器两端瞬时电压降　234
电感器中产生的电势　234
电荷　154, 158
电荷泵　182

电荷分布　158, 162, 222
电荷积累　196
电荷极值　228
电荷量　156
电荷流　176
电荷密度　158
电荷守恒的　154
电荷与时间的关系图　228
电荷运动　164
电荷载体　176, 196
电荷振荡　228
电荷转移过程　172
电解质电容器　170
电介质　168
（静）电力　156, 158
电力的大小　156
电流　176, 220, 230
电流（电压）相量　234, 236
电流（电压）最大值　234
电流表　188
电流产生的磁通量　216
电流超前电压90°　236
电流呈指数衰减　230
电流-电压关系　178
电流方向　176
电流和电压不同相位　234, 236
电流和电压同相位　232
电流环　206
电流极值　228
电流计　188, 214, 226
电流计指针　214
电流密度　176, 178
电流平衡值　226
电流平均值　232
电流随时间的变化率　230
电流稳步下降　220
电流与储存的电荷差90°相位　228
电流与电压间的相位角　238
电流与电压相差90°相位　236
电流与时间　228, 230
电流元　198
电流增大至平衡值　230
电流振荡　228
电流正弦振荡　230
电流滞后电压90°　234
电路　178, 230
电路图　182

电偶极矩 174
电偶极子 174
电排斥力 156
电容 228
电容器 168, 170, 230
电容器充电 230
电容器储存的能量 228, 230
电容器的电场 228
电容器的电压滞后电流90° 238
电容器电容 170
电容器放电 230
电容器两端瞬时电压降 236
电容器之间的电势差（电压） 172
电势差（电压） 166, 178
电势降 184, 186
电通量 160, 224
电位器 188
电吸引力 156
电信 258
电压表 188
电压降 232
电压降低 240
电压升高 240
电子 154
电子流 176
电子气 176
电子器件产生的微波 244
电阻 178, 182, 184, 188
电阻率 178
电阻率呈线性变化 180
电阻率的温度系数 180
电阻率降低 180
电阻率降为零 180
电阻率趋于有限值 180
电阻率随温度的变化 180
电阻率增加 180
电阻器 178, 230, 232
电阻器的电压与电流同相位 238
电阻器两端的电势差 184
电阻器两端的电压降 232
叠加原理 104, 158, 268
定量测量 14
定律 14, 120
定容气体温度计 114
动镜 276
动量 58, 76
动量变化量 60

动量守恒定律 58
动摩擦力 44
动能 54
动能损耗 56
独立方程数 186
杜瓦瓶 126
端电压 182
短程力 130
短路 170
断裂极限 82
对称电荷分布 162
对称面 62
对称物体 62
对称轴 62
对流 126
对数衰减量/率 98
多边形加法法则 24
多重内反射 258
多回路电路 186
多普勒效应 106
多原子理想气体 130

E

二维碰撞 60
二维运动 30
二象性 248

F

发电机 218
发电机频率 218
发动机 146, 148
发散透镜 262
发射体面积 128
法拉第电磁感应定律 220, 224
法线 250
法向力 44
反射 250
反射产生的相位差 272
反射光（线） 250, 252, 276, 290
反射光栅 280
反射角 250, 258
反射面 250
反射起偏 284
反射式望远镜 266
反作用力 50
范艾伦辐射带 212
范德格拉夫起电机 164

方解石　288
方向　22, 28, 34
方向均匀的电场　160
房屋隔热　126
放大的像　260
放大率　266
放电　172
放电电容器　228
非保守力　44, 56
非常光　288
非常遥远的物体　266
非极性分子　174
非晶材料　288
非均匀磁场　192
非理想成像　264
非黏滞流体　90
非欧姆材料　178
非偏振光　290
非偏振（自然）光波　284
非球体高斯面　162
非弹性碰撞　60
非稳恒流　92
非线性波　104
非线性电流-电压关系　178
非线性曲线　140
非相干光源　268
菲涅耳衍射　278
斐索（干涉）技术　248
沸点　114
沸腾　122
分辨能力　280
分隔带电导体的材料　168
分量　22
分流器　188
分子的平动　132
分子的运动　132
分子的振动　132
分子的转动　132
封闭高斯面　162
封闭路径　224
封闭曲面　160, 222
封闭曲面的积分　160, 222
封闭体积　222
夫琅禾费衍射　278
浮力　88
浮体　88
符号相反的点电荷　156

符号相反的电荷　174
符号相同的点电荷　156
辐射　126
辐射形式　244
负电荷　154
负电通量　160
负方向　20
负加速度　36
负焦距　262
负矢量　24
负速度　34
负位移　32
负载电阻　182
附在弹簧上的物体　94
复合波　110
复杂分子　132
傅科电流　220

G

伽马射线　244
改良状态方程　140
盖-吕萨克定律　120
概率　136, 150
概念　14
干扰源　102
干涉　104
干涉和衍射产生的衍射光束　280
干涉条件　268
干涉图样　104, 268, 270
感抗　234
感应电动势　214, 220
感应电动势方向　216
感应电荷　154
感应电流　214, 220
感应电流的产生　214
感应电流方向　216
感应极化　174
感应矩　174
感应偶极子与外磁场方向相反　210
刚体　66
刚体的转动惯量　70
刚体模型　66
钢条　116
杠杆臂　72
高度仪　134
高能量电子减速产生的X射线　244
高能物理学　14

高频电磁波 248
高品质（低品质）电路 238
高斯定律 222
高斯面 162
高温超导体 180
高温热源 142
高温物体或高温分子产生的红外波 244
高压电容器 170
高折射率 258
隔热墙 126
各向异性材料 288
给定面积的线圈 214
工作介质 142
功率 54
功率因数 238
功能定理 54
共点力 64
共焦显微镜 266
共面力 64
共振 100
共振频率 100, 238
共振曲线 100
孤立磁极 204
孤立导体 164
孤立系统 56, 58, 76, 124, 150
固定轴 66
固体的形变 82
固有频率 98, 100
观测者 106
观察者 260
管道 90
惯性 46, 70
惯性参考系 46
惯性定律 46
惯性质量 46
光 248, 290
光（波）的传播 256
光波 284
光波长 274
光波干涉观测方法 274
光程差 268, 270, 272, 280
光的波长 264
光的发散 278
光滑表面 250
光滑路径 90, 92
光路 256
光路图 260

光强 248
光栅 280
光束 248, 250, 258, 276
光束分成两束 276
光速 248, 252
光纤 258
光线 248, 250, 252
光线进入介质 252
光线遇到边界 252
光线在边界处弯曲 252
光学 14
光学透镜测试 274
光源 248
光轴 260, 264, 288
光子 248
光子的能量 248
广义安培定律 200
广义相对论 78
轨道 80
轨道磁矩 206
轨道角动量 206
轨道速度 80
轨道运动 206
轨道周期 80
轨迹 40
轨迹最高点 40
滚动 30, 74
滚动的物体 74
滚动摩擦 74
国际单位制 18
过程 120, 142
过端点的旋转轴 70
过中心的旋转轴 70
过阻尼振动 98

H

海拔 86, 134
海拔增加 134
海平面 134
耗散力 98
合成波函数 104
合力 44
合矢量 24
合外力 64
合外力矩 64, 76
和磁场方向相同的偶极子 210
和子波相切的表面 256

核反应中放射性原子核产生的伽马
　　射线　244
核物理学　14
黑洞　78
黑洞合并　78
黑体　128
恒定（匀）加速度　36
恒定相位关系　268
恒力做工　52
恒温器　116
横波　102
横截面面积　82
横向电磁波　242
横向放大率　260, 266
横坐标　20
红光偏折最小　254
红外波或热波　244
红外线温度计　114
虹膜　266
弧长　68
琥珀　154
互感　226
互感系数　226
华氏温标　114
滑环　218
环流　220
环形线圈　202
黄铜条　116
回复力　44, 96, 98
回路　186, 226
回路定律　186
回旋加速器　192
回旋频率　192
会聚透镜　262
彗差　264
惠斯通电桥　188
活塞　86, 148
火花塞　148
火箭推进　50
霍尔场　196
霍尔电压　196
霍尔系数　196
霍尔效应电流传感器　196

J

机械波　102
机械系统　96

基本定律　156
基本力　44
基本粒子物理学　14
基频　110
畸变　264
极　190
极板间距　170
极板面积　170
极化　174
极限　36
极性　196, 216
极性分子　174
极轴　20
极坐标系　20
几何光学　248
加法交换律　24
加法结合律　24
加速度　36, 48
假力　44
假设　14, 140
假想粒子　62
间隔距离　156
检偏器　286
检验电荷　158
检验电荷带电量　158
减速力　98
剪切　82
剪切角　84
简单放大镜　266
简谐运动　30
简谐振子　94
简正模式　110
箭头　194
箭尾　194
降压变压器　240
交变电动势　218
交变电动势随时间呈正弦变化　218
交流电流　232
交流电路　232
交流电压　240
交流发电机　218, 232
焦点　80, 260
焦距　260, 262
焦面　264
角波数　102
角动量　76, 206
角动量守恒定律　76

角动量随时间的变化率　76
角度　26
角放大率　266
角分辨率　278
角加速度　68
角膜　266
角频率　94, 102, 192
角速度　68
角位移　68
角位置　68
接触点　74
接触力　44
接地导体　154
节点　184, 186
节点定律　186
结冰　122
介电材料　170
介电常数　170
介电强度　170
介质　250, 252
介质分界面　258
介质中的光速　252
介质阻尼系数　98
界面　258
金属　180
紧密缠绕的线匝　202
进气冲程　148
进气阀　148
进入曲面的电通量　160
近似双曲线　140
近视　266
近圆轨道　80
近轴光线　260
浸没的物体　88
经典物理学　14
经典原子模型　206
经验常数　140
晶面　282
晶体材料　288
晶体结构　282
晶体物质　208
晶状体　266
径向加速度　42
净磁场　202
净磁通量　204, 222
净电荷　162, 222
净电通量　160, 162, 222

净力　44, 48
净外力矩　76
静电计　154
静电力　156
静电平衡的导体　164
静电屏蔽或法拉第笼　164
静电起电机　164
静电学　154
静摩擦力　44
静止的检验电荷　158
静止的物体　46
静止粒子　156
静止线圈　216
镜面　250, 260
镜面反射　250
镜面方程　260
镜子　276
酒精温度计　114
居里定律　210
居里温度　208, 210
矩形平板　70
矩形线圈　194
矩阵代数　186
具有永久磁偶极矩的原子
　（离子）210
距离　32
距离的平方　156
-聚焦光　266
绝对零度　114
绝对温度　150
绝对压强　86
绝热过程　120, 146
绝热膨胀　148
绝热压缩　148
绝热自由膨胀　120
绝缘导体　168
绝缘体　154, 168
均方根电流　232
均方根电压　232
均方根速率/方均根速率　136
均分定理　132
均匀磁场　190, 192, 202
均匀电场　160

K

卡诺定理　146
卡诺热机　146

卡诺热机效率　146
卡诺循环　146, 148
开关　226, 230
开关闭合　228
开口压力计　86
开路　230
开路电压　182
开氏温标　114
抗磁　206
抗磁性物质　210
考虑空气阻力的均匀重力场　38
靠近线圈的方向　216
可变电容器　170
可忽略的分子间作用力　140
可忽略的分子体积　140
可见光　212, 244
可见光波长　254
可见火花　172
可逆摆　96
可逆过程　142, 146, 150
可压缩介质　106
刻线（凹槽）　280
空间位置　20
空间针孔　266
空间中的点　20
空气薄膜　274
空气的重量　134
空心圆柱体　70
库仑常数　156
块体金属　220
快速升温　148
宽频带　244

L

拉力　82
拉应变　84
拉应力　84
劳厄图谱　282
棱镜　254
棱镜顶角　254
棱镜对光的偏折　254
棱镜对光的折射　254
棱镜分光镜　254
厘米-克-秒制　18
离开曲面的电通量　160
离散晶面　282
离心力　44

理论　14
理论发动机　146
理想流体　90
理想气体　118, 130, 140
理想气体定律　118
理想吸收体　128
理想系统　98
力　44
力（矩）臂　72, 194
力的冲量　58
力的方向　72
力的作用点　64, 72
力的作用线　64, 72
力矩　72, 174
力矩大小　72
力偶　64, 194
力学　14
力做的功　52
粒子系统的转动惯量　70
粒子行为　248
连接　172, 184
连接导线　182
连接电感器的电容器　228
连接电阻器的电池　182
连接一个电感器的交流电路　234
连接一个电容器的交流电路　236
连接在交流发电机两端的电感器　234
连接在交流发电机两端的电容器　236
连续电荷分布　158
连续性方程　90
量热法　122
量热计　122
两两相互作用　50
两束光重新会聚产生干涉图样　276
两束光传播路径不同　276
亮圆条纹　274
量的变化　22
量子化的电荷　154
量子霍尔效应　196
量子力学　14
列车减速　220
临界电阻　230
临界角　258
临界体积 V_c　140
临界温度　180
临界温度 T_c　140
临界压强 p_c　140

临界阻尼振动 98
零电阻 188
零加速度 36
零拍波 110
零速度 34
零位移 32
流（动） 90, 92
流层 92
流出表面的磁场线数量 204
流出节点的电流 186
流动中的流体 90
流管 90
流入表面的磁场线数量 204
流入节点的电流 186
流体 86, 88, 90
流线 90
流线簇 90
路径 32, 40
路径半径 192
路径长度 32, 110
路径的切线 42
路径积分或线积分 166
路径所围面积 224
路径线积分 224
罗默方法 248
螺线环 202
螺旋轨迹 192
螺旋形式 202
裸眼 266
洛伦兹力 192

M
马赫数 108
马蹄磁铁 190
[麦克斯韦-]玻尔兹曼分布函数 136
麦克斯韦方程组预言的电磁波 242
脉冲直流电流 218
漫反射 250
米 18
密度 86
密狭缝 280
面电荷 174
面电荷密度 158
面积分 204
面积矢量 214
面积为A的平面 160, 204
面偏振光波 284

秒 18
明条纹 268
模糊的像 264
摩擦力 44
摩擦热损失 146
摩尔比热 122
摩尔数 118
摩尔质量 118
末动量 58
末速度 36
某一点的磁场 198
目镜 266

N
南极 190, 212
内禀磁化强度 206
内电场 174
内反射 258
内力 44, 58
内摩擦或黏滞现象 92
内能 124
内阻 182
能 54
能量变化 54
能量来源 182
能量量子化 132
能量守恒 56
能量损耗 56
能量振荡 228
能量转换 56, 228
能量转移 78
逆时针 20
逆时针电流 220
逆时针旋转 68
逆时针运动 30
黏度系数 92
黏性流体 92
凝华 122
凝结 122
凝聚态物理学 14
牛顿 48
牛顿的万有引力理论 38
扭摆 96
扭转的线圈 194

O
欧姆材料 178

偶极子 210

P

帕斯卡定律 88
拍波 110
排斥力 154
排列一致的程度 174
排气冲程 148
排气阀 148
排水体积 88
抛物面镜 266
抛物体 40
抛物线 40
抛物线路径 40
抛物运动 30
膨胀 146
碰撞 60
碰撞次数 138
碰撞粒子 60
碰撞频率 138
偏导数 166
偏光技术 286
偏离平衡位置的位移 94
偏折角 254, 280
偏振波的强度 286
偏振度 284
偏振方向 284
偏振光 290
偏振光波 284
偏振光的强度 286
偏振面 284
偏振片 284
漂移速度 176, 196
频率 94, 102, 244
品质因子 98, 238
平凹透镜 262
平板玻璃表面 274
平的条带 196
平动 30
平动动能 74
平动平衡的条件 64
平方反比（定）律 78
平衡状态 64, 150
平均电流 176
平均功率 54, 238
平均加速度 36
平均角加速度 68

平均距离 138
平均速度 34
平均速率 136
平均体膨胀系数 116
平均线膨胀系数 116
平均自由程 138
平均自由时间 138
平面 160
平面波 102, 256
平面电磁波 242
平面对称 162
平面镜 260
平面内的点 20
平面内的运动 30
平坦表面 250
平条带的上（下）边缘 196
平凸透镜 262, 274
平行板电容器 170
平行的磁偶极矩 208
平行四边形的对角线 24
平行四边形面积 28
平行狭缝 280
平行轴定理 70
坡印亭矢量 242
泊肃叶方程 92
普朗克常数 248
普朗克辐射定律 128

Q

起偏角 290
起偏器 286
起始电荷 228
气态 140
气体 118
气体做的功 124, 148
气压分布 134
气液平衡态 140
汽化潜热 122
汽油发动机 148
千克 18
潜热 122
欠阻尼振动 98
强磁场 202
强度分布 278
强核力 44
强制对流 126
切线 34

切向加速度 42
切向力 84
切应变 84
切应力 84
轻质悬线 96
清晰的像 264
求和 198
球差 264
球对称 162
球面波 102, 256
球面波前 108
球面镜 260
球体高斯面 162
球形电容器 170
球形分子碰撞 138
球形壳 170
球坐标系 20
曲率半径 42, 274
曲率半径（中心） 260
曲率中心 42
曲面覆盖的电荷 162
曲面内电荷 162
曲线路径 40, 42
曲线下的面积 52, 148
曲线斜率 34
曲线运动 30
驱动力 100
全反射 258
全反射光束 258
全息术 290
全息影像 290

R

燃烧 148
扰动 102
绕过障碍物的波 278
绕在铁芯上的线圈 240
热 114, 122
热泵 144
热传导 126
热传导定律 126
热导率 126
热机 142, 144, 146
热寂 150
热接触 114
热力学 14
热力学变量 118

热力学第二定律 142
热力学第零定律 114
热力学温标 114
热能 122, 142, 144
热能量 150
热能吸收 142
热能转换 122
热膨胀缝 116
热平衡 114
热容 122
热损失 126, 146
热效率 144
任意点的力矩 194
任意方向 250
容抗 236
熔化 122
熔化潜热 122
入射光（线） 250, 252, 290
入射角 250, 258
入射面 250
入射在检偏器上的波 286
软铁磁体 208
软铁芯 240
瑞利判据 278
弱磁场 202
弱核力 44

S

三角形的边 24
三角形加法法则 24
三维衍射光栅 282
三相点 114
散射起偏 284
色差 264
商用发电机 218
熵变 150
熵函数 150
熵函数减少 150
熵函数增加 150
上层晶面 282
射线近似 248
摄氏温标 114
升华 122
升温 148
升压变压器 240
声波 106
声波速度 106

剩余磁化强度 208
剩余气体 148
失重 38
时间标准 18
时间常数 230
时间函数 32
时间间隔 32, 36
实像 260
实心球 70
实心圆柱体 70
实验 14
实验观测 14
实验曲线 140
矢量 22, 34
矢量构成的平面 28
矢量和标量相乘 24
矢量积 22, 28
矢量加法 24
矢量减法 24
矢量起点 22
矢量相等 24
矢量终点 22
（电）势 166
势的叠加 166
势能 54, 166
视杆细胞 266
视网膜 266
视锥细胞 266
室温气体 118
释放的热能 144
受迫振子 100
输出电压 240
数值 22
数字万用表 188
双凹透镜 262
双缝干涉图样的强度分布 270
双金属条 116
双曲线 120
双凸透镜 262
双原子分子 132
双折射材料 288
双折射率 288
双折射起偏 284
水/水蒸气/冰 114
水的冰点 114
水的汽点 114

水平方向 40
水平距离 40
水平射向镜子的光线 276
水平位置 40
水银气压计 86
水银温度计 114
水银柱 86
顺磁 206
顺磁性物质 210
顺时针 20
顺时针电流 220
顺时针旋转 68
顺时针运动 30
瞬时电流 176
瞬时电压降 232
瞬时功率 54
瞬时加速度 36
瞬时角加速度 68
瞬时速度 34
斯蒂芬-玻尔兹曼常数 128
斯蒂芬-玻尔兹曼定律 128
斯涅尔定律 252
斯特林发动机 48
四冲程循环 148
速度 34
速度的垂直分量 40
速度的水平分量 40
速度平方的平均值 130
速度-时间图 36
速度矢量方向变化 42
速度梯度 92
速度选择器 192
速率 36
速率变化 42
速率分布 136
速率分布函数 136
速率分布曲线 136
速率分布曲线达到峰值 136
速率在v到$v+dv$区间的分子 136
随机排列的磁畴 208
随机取向的电偶极子 174
随机取向的偶极子 210
随机游走过程 138
随机运动 30
随遇平衡 64
缩小的像 260

T

太阳　80
太阳产生的紫外线　244
太阳常数　128
弹簧摆　96
弹簧做功　52
弹性　82
弹性摆　96
弹性极限　82
弹性模量　82
弹性碰撞　60
弹性势能　54
碳刷　218
逃逸速度　80
梯度算符　166
体电荷密度　158
体积　118
体积变化　116
体积密度　70
体积模量　84
体积弹性　84
体积通量（流量）　90
体积应变　84
体积应力　84
填充介质的电容器　170
条纹　268
条纹级数　268
条形磁铁　190
铁磁　206
铁磁体　208
铁磁性物质　208
铁屑　190
通大气　86
通电导体　190, 194, 198
通电导线　194
通电的　154
通电线圈　194
通电线圈上的力矩　194
通过闭合表面的磁通量　204
通过封闭曲面的净电通量　162
通过面积元的磁通量　204
通过曲面的电通量　160
通过线圈的磁通量　216, 226
通过一个线圈的磁通量穿过另一个
　　线圈　240
通量变化率　224
通用气体常数　118

同心　170
同性电荷　154
同一平面　250
同轴　170
瞳孔　266
投影　22
透镜　262
透镜表面的曲率半径　262
透镜成像公式　262
透镜的焦强　266
透镜系统　262
透镜中心　264
透明棒　258
透明介质　252
透明狭缝　280
透射光的强度　286
透射光栅　280
透射（偏振）轴　286
凸凹透镜　262
凸面镜　260
图像斜率　36
湍流　92
推力　50
退磁物质　208
陀螺　76
陀螺仪　76
椭圆轨道　80
椭圆偏振光波　284

W

外部磁通量　216
外部电荷　158
外电场　174
外加电压呈正弦变化　238
外力　44, 46, 58, 100
完全发射体　126
完全非弹性碰撞　60
完全浸没的物体　88
完全偏振光　290
完全弹性碰撞　60
完全吸收体　126
万有引力　38
万有引力常量　78
望远镜　266
微波　244
微观模型　130
微观区域　208

微粒（粒子）理论 248
微小的细节 266
围绕质心的转动 66
维恩位移定律 128
未充电的电容器 172, 230
未磁化物质 208
位移 32, 52, 102
位移方向 52
位移精准测量 276
位移矢量 32
位移振幅 106
位置 20, 32
位置变化 32
位置-时间图 32, 34
位置矢量 32
温标 114
温度 118
温度变化 116
温度从T_1降到T_2 148
温度计 114
温度梯度 126
稳恒磁场 190
稳恒电流 176, 198, 200, 218
稳恒流 90
稳态电路 230
稳态条件 100
涡流 220
无电阻的LC电路 228
无滑动的滚动 74
无介质电场 174
无量纲数 252
无摩擦的滑轮 48
无穷大电阻 188
无限小过程 150
无线电波 244
无相位差 272
无序 150
无旋流 92
无质量的轻绳 48
无阻尼振子 98
物（像）方焦点 262
物镜 266
物距 260
物理（复）摆 96
物理接触 60
物理量 14
物理现象 14

物理学 14
物体 30
物质 208, 210
物质的磁化强度 206

X

吸收的热能 144
吸引力 78, 154
希腊字母德尔塔Δ 32
稀疏（紧密）螺线管 202
系统的初始状态 124
系统的宏观状态 124
系统的微观状态 124
系统的质心 62
系统的中间状态 124
系统的状态 124
系统的总质量 62
系统的最终状态 124
细长杆 70
狭缝 280
狭缝间距 280
下层晶面 282
纤维光学 258
现代物理学 14
线电荷密度 158
线积分 200, 224
线密度 70
线偏振电磁波 242
线偏振光波 284
线圈 202, 214, 216
线圈的电感 226
线圈逆时针旋转 194
线圈平面 194
线圈顺时针旋转 194
线性波 104
线性电流-电压关系 178
线性动量 58
相对加速度 36
相对论 14
相对论动能 54
相对速度 34
相对于原点的位置矢量 76
相对运动 30
相干光源 268
相干光源对 270
相隔一段距离的电荷 174
相消干涉 104

相消干涉条件 268
相长干涉 104
相长干涉条件 268
响亮的爆炸声 108
向上的力 86
向下的力 86
向下滚动 74
向心加速度 42, 192
向心力 42, 44
向一个方向偏转 214
相变 122
相量 232
相量图 232, 234, 236
相位 94
相位常量（相位角） 94, 102, 104
像 260, 262, 264
像差 264
像场弯曲 264
像距 260
像散 264
橡胶或玻璃棒 154
小磁化率 210
斜率 32
斜面 48
谐振腔 128
行波 102, 104
行星运动 80
形状不规则的导体 164
形状弹性 84
虚像 260
悬线长度 96
旋涡区 92
旋转 68, 192
旋转的带电粒子 192
旋转木马 76
旋转趋势 72
旋转中心 42
旋转周期 192
旋转轴 66, 72, 218
选择性吸收起偏 284
寻常光 288
循环 142, 148
循环过程 124, 142
循环效率 148

Y

压力随深度变化 86

压强 86, 118
压强变化 106, 108
压强幅度 106
压强随海拔的变化 134
压缩 82, 146
压缩冲程 148
压缩区域（密部） 106
压缩性 82
亚声速 108
沿电动势反方向被穿过的电动势 186
沿电动势方向被穿过的电动势 186
沿电流反方向被穿过的电阻器 186
沿电流方向被穿过的电阻器 186
沿径向向外 158
沿直线方向 156
衍射 278
衍射光束 280
衍射光束互相干涉 280
衍射光束形成最终图样 280
衍射光栅 280
衍射角 278
衍射图样 278
眼房水 266
验电器 154
杨氏模量 84
液氮 180
液化气体 140
液态 140
液体 86
液压 88
液压车桥 88
一次（二次/……）谐波 110
一段导线 194
一段任意形状的导线 194
一段直导线 194
一对点电荷的电势 166
一级点源或振子 256
一维运动 30
一系列颜色 254
一组电荷 158
移动的带电粒子 190, 192
以导线为中心的圆形路径 200
以度为单位的角度 68
以弧度为单位的角度 68
以焦耳热形式耗散的能量 228
异性电荷 154
阴影区 278

音爆 108
引力 38, 78
引力波 78
引力场 78
引力辐射 78
引力质量 46
英尺-磅-秒制 18
应变 82, 84
应力 82, 84
应力应变比 82
硬铁磁体 208
永磁体 190
永久磁偶极矩 208
永久极化 174
用丝绸或毛皮摩擦 154
由闭合路径决定的平面 200
由多个电荷引起的力 156
油气混合物 148
有两条相互平行狭缝的屏幕 270
有限尺寸螺线管 202
有限大的导体 198
有效直径 138
有心力 44
有阻尼的简谐运动 230
右边线圈与负载电阻相连 240
右手定则 28, 192, 200
余弦 26
余弦函数 24
与磁场平行的磁畴 208
与路径无关 166
宇宙射线 212
原点 20
原子和分子内电子重新排列产生的可见光 244
原子间距 282
原子物理学 14
原子阵列 282
圆的半径 68
圆弧 66
圆弧的半径 68
圆环 70, 202
圆孔 278
圆盘 70
圆条纹 274
圆条纹半径 274
圆形轨道 80, 206
圆形轨道半径 206

圆形轨道面积 206
圆周 68
圆周长 200
圆周路径（轨迹） 42, 192, 200
圆周运动 30, 66
圆柱壳 70, 170
圆柱形电容器 170
远视 266
月球 80
匀速 34
运动 30
运动定律 46
运动（的）物体 30, 46
运动学 30

Z

在 v 到 $v+dv$ 区间 136
在磁场中运动 220
在粗糙表面上滚动 74
在轨运行的电子 206
载流子 196
载流子符号 196
载流子密度 196
张力 44
折射光（线） 252, 290
折射角 252, 258
折射率 252, 258, 264
折射率随波长的增加而减小 254
折射式望远镜 266
真空 126
真空磁导率 198, 224
真空介电常数 156, 162, 222, 224
真空中的光速 252
真实流体 90
真实（实际）气体 130, 140
真实系统 98
振荡 30
振荡电荷产生的电磁波 242
振荡角频率 228
振荡周期 234
振动 30, 94, 98
振动次数 94
振动模式 110
振动曲线包络 98
振幅 94, 102
蒸发 122
蒸汽机 148

整流器　218
正磁化率　210
正电荷　154
正电通量　160
正方向　20
正功　100
正极　188
正加速度　36
正检验电荷　158
正焦距　262
正碰撞　60
正速度　34
正位移　32
正弦波　102
正弦函数　24
正像　260
支点　72, 96
直导线　198
直角坐标系　20
直流电流　218
直流发电机　218
直线传播　278
直线上的点　20
直线运动　30
指南针　190
指数型　134
指数因子　98
指向电荷　158
指向纸面外（里）的磁矩　194
制动系统　220
制冷机　144
制热系数　144
质点　30, 90, 138
质点的角动量　76
质点模型　30
质点系　62
质点系的角动量　76
质点系的牛顿第二定律　62
质点系的总动量　62
质量　48
质量标准　18
质谱仪　192
质心　74, 96
质心的加速度　62
质心的平动　66, 74
质心的矢量位置　62
质心的速度　62

质心的位置　62
质心的坐标　62
中间最厚的镜片　262
中心引力　78
重力　38
重力（引力）场　38
重力加速度　38
重力势能　54
重量　38, 46, 78
重心　62
周期　94, 102
周期力　100
周期（振荡）运动　94
周围环境　126
主极大　280
主轴　260
驻波的波函数　110
柱对称　162
柱坐标系　20
转动　30
转动的运动学方程　66
转动动能　74
转动惯量　70
转动能　70
转动平衡的条件　64
状态方程式　118
状态函数　124, 150
锥　108
锥表面　108
锥面波前　108
子波　256
紫光偏折最大　254
紫外线　244
自动调焦　266
自感　226
自感电动势　226
自然对流　126
自行产生的磁通量　216
自旋　206
自旋磁矩　206
自旋角动量　206
自由电子　164
自由度　132
自由度的个数　132
自由落体　38
自由落体加速度　38
自由落体运动　38

自由物体受力图　48
字母上方的箭头　22
总磁场　198
总磁通量　204
总电荷　172
总电流　200
总电通量　222
总动量　58, 60
总动能　74
总功　52
总机械能　54
总加速度　42
纵波　102
纵坐标　20
阻抗　238
阻力　44, 220
阻尼率　98
阻尼拍波　110
阻尼系数（阻尼比）ζ　98, 100
阻尼系统　100
阻尼振子　98
最大电流　232
最大高度　40
最概然速率　136
最高效热机　146
最佳反射面　258
最小强度条件　278
最终长度　116
最终平衡状态　150
最终位置　32

左边线圈与交流输入电压源相连　240
作用力　50
作用力和反作用力定律　50
作用于物体的力　52
坐标　20
坐标系　20
坐标轴　20
坐标轴的名称　20
坐标轴集　20
做的净功　142
做功冲程　148

数字和西文字母开头的词语

1/4周期　110
180°相位差　272
BCS理论　180
$B\text{-}H$曲线　208
D形盒　192
LC电路　228
LIGO激光干涉引力波天文台　276
N匝线圈　226
RLC串联电路　230
RL串联电路　230
SI——国际单位制　18
X射线　244, 282
X射线束　282
X轴　20
Y轴　20
Z轴　26

Compiled and illustrated by
Iryna Moroz

编撰与图示
伊莉娜·莫拉兹

Edited by
A. T. Augousti
(United Kingdom)
Yang Fuling
Liu Tianyu
(P. R. China)

主编
A. T. 奥古斯汀
（英国）
杨福玲
柳天宇
（中国）

Translated by
Chen Congcong
Tan Lu
Zhang Qingluan
(P. R. China)

译
陈聪聪
谭　露
张青滦
（中国）